# BOTANY AT THE BAR

The Art and Science of Making Bitters

# BOTANY AT THE BAR

## The Art and Science of Making Bitters

Bring the flavor and stories of plants from around the world
to your beverage through science-based craft bitters.

Selena Ahmed • Ashley DuVal • Rachel Meyer
Creators of Shoots & Roots Bitters

Cocktail Recipes: Christian Schaal, Kevin Denton, and Shoots & Roots
Shrub Recipes: Jim Merson

*Leaping Hare Press*

First published in the UK in 2019 by
*Leaping Hare Press*
An imprint of The Quarto Group
The Old Brewery, 6 Blundell Street, London N7 9BH, United Kingdom
www.QuartoKnows.com

British Library Cataloguing-in-Publication Data
A catalogue record for this book is available from the British Library

ISBN: 978-1-78240-560-3

This book was conceived, designed, and produced by
*Leaping Hare Press*
58 West Street, Brighton BN1 2RA, United Kingdom

*Publisher* Susan Kelly
*Creative Director* Michael Whitehead
*Editorial Director* Tom Kitch
*Commissioning Editor* Monica Perdoni
*Project Editor* Jenny Campbell
*Editors* Imogen Palmer, Theresa Bebbington, and Megan Kaye
*Design* Studio Noel
*Illustrations* Clare Owen
*Photography* Stephen DeVries (pages 7–8) and Xavier Buendia (pages 60–63)

Printed in China

10 9 8 7 6 5 4 3 2 1

# Contents

# Introduction

Humans have always depended on plants.
Ethnobotanists explore this interaction of plants
and people. Many are scientists who explore
traditional uses of biodiversity for medicines,
food and beverages, and other cultural activities.
We are three ethnobotanists who have always been
captivated by the scents, sights, and tastes of the
plants and landscapes where we do our fieldwork.
United by a shared passion to communicate what
we have learned, the three of us came together as
Shoots & Roots Bitters. Here, we create and share
science-based craft bitters that draw from ecology,
practices, and flavors from around the world.

We met during graduate studies more than a decade ago through The New York Botanical Garden. While each of us had different research directions, we all shared a passion for biodiversity and culturally important plants—from studying them to tasting them (when edible). From examining the genetics of wild tea trees deep in the forests of Southwest China to studying historical patterns in crop domestication and the science of flavor, we began to collaborate on projects.

Biodiversity conservation is a critical need in the age of the Anthropocene, as is the preservation of associated traditional knowledge and practices. Scientists are more motivated than ever to communicate these challenges.

Our own efforts to connect with the public would eventually become Shoots and Roots Bitters, a company through which we are devoted to the science of plants and people. This book brings together ethnobotany and recipes about hundreds of important plants from around the world. We hope that our readers can take part in global traditions and botanical diversity in their own homes through the creation of bitters and other botanical beverages.

Why bitters? Bitters are botanical infusions from multiple plant species that, when combined, form a complementary flavor profile. By creating bitters, we are able to tell complex stories, be it their symbology or ecology.

**Below** Shoots & Roots Bitters founders (from left) Rachel Meyer, Selena Ahmed, and Ashley DuVal share the charms of the lab and seek to make research exciting and inclusive.

## A Kitchen in the Lab

Brewing bitters is a creative art and, as scientists, we have a great asset: laboratory equipment. Working in the phytochemistry lab at the City University of New York (phytochemistry being the science of chemicals derived from plants), we were able to play with all sorts of equipment to explore the range of flavors in every plant. It wasn't long until we began using bitters as interactive tools in ethnobotanical workshops, and eventually, we began to produce them commercially.

## What We Do

The mission of Shoots & Roots Bitters is to promote botanical education and conservation through science-based craft bitters and associated events. We host interactive workshops on botanical topics that range from exploring the histories of domesticated plants, scientific approaches toward making and using bitters, the science of flavor, and environmental diversity and sustainability. Coming face to face with plants, getting messy and rowdy extracting them, and tasting these extractions with strangers is a really fun way to spread the word about the achievements of traditional knowledge and botanical research breakthroughs.

While running Shoots & Roots Bitters, all of us have also maintained full-time careers as academic researchers. Meeting so many different people through our company has helped us see science with fresh eyes: it is humbling to discover the depth and variety of scientific knowledge outside of "traditional science." Our experiences have also fired us up as activists. Working together as individuals from different backgrounds, and as women collaborating in a male-dominated industry, has sharpened our awareness of just how important it is to consider multiple perspectives from diverse geographies and sectors of society. We work with as many allies as we can to promote botany in the beverage industry.

In many of the bitters described in this book, our approach is largely place-based, with many of our blends reflecting a diversity of regional plants. Attention to place-based biodiversity is more relevant than ever today because of rapid environmental and social change.

Our quest is to spread understanding and help preserve these irreplaceable aspects of our shared history. New markets and supply chains can be encouraged through connections between science research networks, communities, and the craft innovators we hold as mentors—the chefs, bartenders, consumers, and liquor makers. Forgotten plants may get some time in the spotlight. And yes, that may be in the form of a better martini.

## How to Use This Book
This book covers the cultural, botanical, sensory, and health aspects of bitters; empowers you to make your own science-based bitters; and gives you inspiration for your own drink recipes. Chapter One follows the cultural origins of bitters from ancient times to modern-day cocktail culture. Chapter Two presents the basics of the botany and ecology of bitters to help you better source plants and create your own bitters. Chapter

Three provides an overview on the science of flavor, which will help you understand your own responses to flavor and plant chemistry 101. Chapter Four teaches you how to make bitters for yourself. We start with some basic recipes to get your feet wet with easy-to-access ingredients before moving on to more advanced recipes that use ingredients from around the world. Chapter Five integrates botany with the bar, written in full collaboration with the professionals: Christian Schaal, bar manager and mixologist at Zebulon in Los Angeles; Kevin Denton, Head of Education and Mixology for Pernod Ricard in New York; and Jim Merson, owner of Hot Dog Sodas, who has designed shrubs for some of the freshest joints. Each drink features a special plant profile with science from ethnobotany to natural history. The final chapter contains a plant directory with a broad geographic representation of our 125 favorite botanicals that we have tried and tested. This directory includes many underutilized plants that could benefit from expanded interest in their cultivation, conservation, and use. So you can continue your exploration, at the end of the book you will find advice for making completely new bitters recipes from botanicals you source. We also include some more references for further reading, and online sources for obtaining quality dry ingredients for making bitters.

We hope you enjoy this book as much as we have enjoyed creating it. Raise a glass to people, plants, and the drinkable science in-between.

**Selena Ahmed, Ashley DuVal, and Rachel Meyer**

Chapter 1
# A Short History of Bitters

Has a family member or friend ever give you Swedish bitters and soda for heartburn, or a cup of Jamaican rice bitters to treat congestion? If so, you probably learned from personal experience that many healthcare practices around the world rely on bitter-tasting botanicals. Bitters have abundant flavors, uses, and definitions, which make them a valuable ingredient to have at your fingertips. In this chapter, we will take a look at what bitters are and their origins, cultural diversity, and uses. We hope it will encourage your own experiments, whether for health or for some delicious fun.

# What Are Bitters?

For more than a thousand years, concentrated extractions of plants—bitters—have been used to support health and well-being, to add flavor, and to stimulate the senses all over the world. The earliest use of bitters was for medicine to treat a wide range of ailments from digestive issues to the common cold. By the early 1800s, the use of bitters evolved to become a key ingredient of cocktails.

Every plant has its own unique properties and intriguing flavors, and their preparation varies widely. Botanical bitters can involve using bark, roots, seeds, flowers, fruits, or even the entire plant. The results can be consumed on their own as a tisane (hot water infusion), as a tincture (a concentrated single plant extract), or added to cocktails and culinary innovations.

Most often, bitters are prepared by infusing botanical material in a fermented base, such as grain alcohol, fruit wine, or beer, which extracts, concentrates, and preserves the plants' properties. Other methods of producing bitters include steam distillation of the essential oils in plants or hot water extraction. Each method produces distinct results, which means that the final scent, flavor, and stability of those qualities will vary according to how they were prepared.

Ultimately, the creation of bitters serves to pull out flavor and therapeutic compounds—known as phytochemicals or secondary metabolites—from botanical material into a liquid. While we can benefit from the taste or the physical effects the phytochemicals provide, plants produce them for different reasons: for protection and communication.

As the name suggests, most of these creations have a bitter taste. This, however, is an oversimplification, because they can also contain elements that are sweet, sour, mineral, and umami (savory). They contain hundreds of aromatic compounds that can direct the sensory experience of a drink.

It's time for the term "bitters" to be decoupled from the branding that has shaped our ideals and be redefined in a global context. Bitters reflect identity, stories, and their local botanical cornucopia. Call them what you want—elixirs, tonics, mixed extracts, digestifs, herbal liqueur, vermouths, or amaros—here, we're analyzing the world's botany at the bar through bitters.

# An International Tradition

Before written history, people around the world were trading plants for infusing into beverages. Grains, succulents, fruit, herbs, and honey were the building blocks of innovation for mixed plant extracts that served a wide variety of medicinal, social, and ritualistic purposes. Around the world, the evolution of human societies and culture correlated with feasting and fermented beverages.

## Africa

Many regions in Africa have a rich bitters culture. The African ancestor of all humankind adapted to processing alcohol in the human liver nearly two million years ago; it is thanks to them that most humans can continue to enjoy the products of fermentation. For the bitters enthusiast, Africa is the holy grail, with everything under the sun from a rich archaeology revealing ancient bitters recipes to roadside bitters stands and bars serving local palm wine distillate poured over a mixture of shoots and roots.

**Ancient Egypt:** Ancient Egypt possessed a treasure trove of plant uses—including fermented beverages and bitters—and many of today's common foods had surprising applications at the time. For example, onion skins were used for covering the eyes during embalming. Ancient Egyptians had a refined system of preparing herbal infusions in grape wine that dates back more than five thousand

years. Scientists have discovered herbs, like blue tansy and coriander through chemical analysis of artifacts, which has helped them trace these ancient recipes back farther than the written record. Archaeologists have identified a mid-first millennium recipe from ancient Egypt known as *kyphi* which was used as an incense or mixed with wine and honey as a bitters beverage. Tree resins, including myrrh, fir, terebinth, and pine, were also used in ancient Egyptian medicinal and alcoholic formulas, and are still popular today.

**West Africa:** Many people of this region consume bitters on a daily basis, regarding them as cleansing restorers that can bring their bodies more energy and strength. The persistence of bitters in the Americas, especially around the Caribbean, the southern United States, and Brazil, is directly related to the cultural importance of bitters in West Africa. Ethnobotanist Tinde van Andel documented hundreds of species used for

bitters in Suriname by people of West African descent. The translocation of plants from the Americas to West Africa is also evident—in monasteries in Togo, you'll sometimes find avocado leaf vermouth, nodding to Aztec traditional knowledge.

**Southern Africa:** Boasting the third most biodiverse region of the world and more than five thousand medicinal plants, the arid landscapes of southern Africa are rich in medicinal, resinous, and aromatic plants. Aloe bitters were developed by the San and may be among the most ancient bitters on record, as well as the first bitters exported to Europe. Devil's claw tubers (*Harpagophytum*) or *Terminalia* gum provide medicinal bitters. Melons and bulbs are insanely bitter, sweet, and aromatic, with uses spanning perfumes, food, and digestive tonics.

## THE NEW ANTIQUITY

While many ancient recipes have been lost to time, today there is a broad international interest in saving these botanical windows into our past. Ethnobotanists continue to collect traditional recipes and compare them with the earliest written records. The literature on ancient recipes is particularly rich for beverages in the Americas, Africa, and Asia, Unfortunately for amaro enthusiasts, however, European beverage recipes are relatively scarce. Perhaps the tradition of keeping recipes like Chartreuse and Campari a secret is much older than we thought. Recreating historic recipes from around the world is possible thanks largely to the Internet—it's surprisingly easy to discover the identity of different saps that were used in an original recipe and to source the ingredients from their original country. If you are planning to use plant ingredients that you have identified from an old recipe, keep in mind that today those same plants may be different because of changing environmental and genetic factors. Also be mindful of if the plants are now endangered.

### East and South Asia

From the drawers of dried medicinal plants in traditional Chinese medicine (TCM) to the therapeutic plants used in Ayurvedic cooking, East and South Asia have a rich history and culture of preparing bitter infusions to support well-being in many healing systems.

China: Vessels in Chinese tombs have provided evidence of surprisingly similar recipes to those found in Egypt. Evidence suggests that the earliest stored alcohol in China was a mixed grain-and-fruit mead with herbs dating back more than nine thousand years. Scholars believe that the grain used

**Above** This illustration from the Ming period (1368–1644) shows the prepartion of chrysanthemum liquor, said to warm yin and raise yang, and grape wine, which was thought to replenish chi.

for this alcoholic infusion was rice and millet and that the fruit was hawthorn (which contributes acidity much like wine grapes). The flavors of these early infusions came from wild wormwoods, chrysanthemums, and basils. These ingredients remain important in the hundreds of botanical blends currently used in contemporary Chinese alcohols, known as *jiu*, which offer dessertlike, vegetal, or extremely bitter medicinal experiences.

Traditional Chinese medicine distinguishes functional entities such as chi/qi (energy) or xue (blood), where imbalances such as too much yin (storing), and yang (propulsive) can be treated with bitters. The greater function of bitter tonics in TCM, however, is prevention instead of cure.

Other Chinese indigenous medicinal systems have an extensive history of using fermented beverages with botanicals, both for public events and to bond with friends and family during rituals that include engagements, funerals, and ancestor worship. Unfermented, steeped bitters beverages are also cultural keystones, such as yak butter tea, infused with different herbs and smoky tea, depending on the region.

**India:** The immense Himalayas offer not only an awe-inspiring sight, but an intriguing location for medicinal and aromatic plants. Growing conditions there are challenging, and, during fieldwork, several of our hosts told us that plants from higher ground make the most nutritious, efficacious, and tasty bitters. This is because high ultraviolet stress at high elevation requires the plants to produce more protective antioxidants.

India is home to seven regions, each with their own unique biodiversity. Local flora are often transformed into bitter tonics by juicing, boiling, cold infusions, or extractions in fat, following heritage Ayurvedic principles. Of India's forty-five thousand or so flowering plant species, more than three thousand are documented for their therapeutic potential, including for bitter tonics—an exceptionally high proportion.

## Mediterranean

At about the same time as early botanical infusions were being buried in tombs in ancient Egypt, wine and beer were being used to extract botanicals around the Mediterranean. Ancient Greeks are thought to have taken quickly to viticulture (wine production) once trade with ancient Egypt introduced them to the methods. They then spread wine-making and associated practices of botanical infusions to surrounding countries through trade.

The Middle East is where humans first domesticated plants and grain crops. This process was crucial for the transition from a hunter-gatherer lifestyle to settled agriculture. With the birth of farming came an abundance of grain, which led to the use of grain alcohol for fermentation.

**Mesopotamia:** Along with ancient Egypt, ancient Mesopotamia (now modern Iraq) produced some of the world's earliest recordings of botanical medical treatments. Renowned for its botanical and brewing cultures, it is here that the fertile plains between the Tigris River and Euphrates River boasted beloved tree species, such as walnut, almond, and apricot.

Alcohol brewing took the form of fermenting date fruit syrup with herbs and honey in a Sumerian ale. The climate was inviting, meaning it could support plants from a similar latitudinal range in Europe, north Africa, and Asia. Trade allowed ancient Mesopotamians to take advantage of their favorable setting, and the result was an extraordinarily culturally and biologically diverse cuisine and beverage. The possibilities for concocting bitters here seemed to be endless, and herbal remedies and tonics of Mesopotamia addressed both the physical and spiritual components of disease.

Two classes of practitioners—the *asu*, who were trained in therapeutic medicine, and the *ashipu*, who practiced divinatory medicine—employed bitter infusions both for drinking and rubbing into the skin. Many of these Babylonian bitters are still used in Iraq today.

**Ancient Greece:** Ancient Greeks enhanced the flavor of grape wine by infusing it with resins, herbs, spices, oil, perfume, seawater, and brine (a salt solution). This botanically infused wine served many purposes in ancient Greece, including the economic, the social, the religious, and the medical. Medicinally, botanically infused wine was used as a digestive aid, tonic, and painkiller. These flavor-enhancing practices inspired modern-day retsina, mulled wine, and vermouth.

The ancient Greeks were known for their delicious infused wines, but their elixirs with bitter herbs were considered to have near-magical properties. The term *elixir* has its roots in the Greek word *xerion* meaning "healing powder," used for life-prolonging properties or immortality. In the alchemical text *Isis to Horus*, the prophetess Isis gives her son an elixir referred to as the "drug of the widow" for immortality. Isis herself represents an elixir in ancient mythology as the "water of life" and "dew" that heals and unites the dismembered parts. The right botanicals were, in themselves, a spiritual symbol.

**Italy:** The Etruscans relied on eleven plants in their herbal cabinets for healing, magical, and religious practice. The most central plant, gentian, remains dominant even today in Italian culture.

Gentian is still one of the most common plants found in modern bitters and is almost certain to be a foundational ingredient in alcohols served during the Italian apertivo hour, such as in amaros (translated as "bitter"). Gentian's fame rests in the fact that it contains one of the most bitter compounds known to human kind—amarogentin. Valued for stimulating the liver and appetite, gentian is also used to aid digestion, increase red blood cell production, and boost the immune system.

## DOCTRINE OF SIGNATURES

Out of the estimated four hundred thousand plant species on this planet, how do we know which ones are therapeutic? One theory is that humans observed the plants that animals consumed and experimented with the edible and medicinal properties of these plants based on their observations. Another theory is that of the Doctrine of Signatures.

Dating back more than two thousand years to the Greek pharmacologist Dioscorides (40–90 C.E.), the Doctrine of Signatures holds that plant parts with a similar shape to human organs may be suitable for curing problems with those organs—for example, consuming walnuts or macambo beans to strengthen brain function. While this isn't exactly hard science, we do know that many plants chosen according to the Doctrine of Signatures involve bitter botanicals, so it may at least, however accidentally, have pointed people in the right direction.

## South America

The thick jungles of the Amazon, with lianas, hallucinogenic plants, and shamans, are probably the images most frequently used to represent medicinal plants of South America, and many of the bitter plants continue to mark cultures there to this day. The Peruvian flag even includes two medicinal bitter plants from the Amazon forest: cinchona (quina; *Cinchona* spp.) and sarrapia (*Dipteryx odorata*). Both were first used by various Amazonian ethnic groups and eventually became two of the key ingredients of Amargo Chuncho, or Peruvian Chuncho Bitters.

The bark of cinchona trees is the source of one of the most well-known bitter compounds from South America: quinine. The Quechua Amerindians indigenous to Peru, Bolivia, and Ecuador prepared a tonic water with the ground bark of cinchona trees as a muscle relaxant. The Jesuits learned about the powers of cinchona during their work in South

**Above** Cinchona bark has been harvested in the Andes for centuries; its value was such that Colombia, Ecuador, and Peru banned the export of cinchona seeds in the mid-nineteenth century.

America and introduced dried cinchona bark to Europe in the sixteenth century. Initially used in Europe to treat diarrhea, by the seventeenth century, it was used to treat malaria.

## Mexico and Central America

When the Spanish conquistador Hernando Cortez encountered the Aztecs in modern-day Mexico, he was offered a bitter-tasting drink the color of blood. The last reigning Aztec emperor, Montezuma II, was said to consume at least fifty servings of this drink daily. This drink was chocolate, prepared as a bitter infusion. This was the first time Europeans had encountered chocolate, made from the dried and fermented roasted beans of *Theobroma cacao* (meaning "cacao, food of the gods").

Although native to the Amazon, chocolate, or cacao, moved north into meso-America through early trade routes and was domesticated along the way. Indian corn, beer, and cacao became central components of feasting culture in Mezoamerica. Cacao was also blended with other botanicals to create both medicinal and recreational drinks. The Maya created a cacao porridge that they thickened with cornmeal, sweetened with honey, and decorated with achiote and chili peppers for a complex floral and spicy flavor.

## The Caribbean

The Caribbean islands have produced a myriad of bitters, including those made of rice bitters, cerasee (*Momordica charantia*), and aloe vera. Rice bitters are not actually made of rice but from the bitter herb *Andrographis paniculata*, also called the king of bitters, which is thought to have been brought into Jamaica by Vietnamese refugees.

**Above** Made without sugar, Mayan chocolate has a strong bitter taste. The word *chocolate* is derived from Mayan name for it, *xocolatl*, which means "bitter water."

Many of the Caribbean bitters are prepared as bush teas and have become one of the most socially accepted forms of medicine in the region, forming part of daily rituals and consumed by people of all ages. The Caribbean enjoys a traditional medicinal system rife with spiritual stories and rituals and bitters form a central part, meaning that consuming them is a cultural norm. Many of the herbs used in bitters are not only ingested but used in the form of bitters baths to treat the body externally.

## Oceania

Oceania has a fascinating history of bitter medicinal tonics and fermentation developed by Aboriginal peoples, drawing on the

## KAVA

The South Pacific is the home of another frothy bitter tisane that has helped nurture social relationships in ceremonious meetings, often between feuding tribes. This is *kava*, meaning "bitter" in Polynesian. Unlike other bitters, kava was not traditionally extracted in alcohol and initially was not even extracted in water. Instead, it was chewed, spat out, strained, and drunk via a shared bowl.

Fortunately for the more squeamish, a good soak in water works well, too. Kava infusions are valued for relaxing the muscles, particularly in the neck and shoulders, without fogging the mind. These properties are becoming more widely recognized and increasingly being served in coffee shops and alcohol-free bars; kava should never be taken along with alcohol.

incredible floras spanning tropical, rainforest, marine, desert, and high mountain ecosystems. Early on, sugar-containing plant parts, such as banksia flowers, gum tree sap, quandong roots, and pandan nuts, were set in water to ferment. Researchers are now chasing these recipes; their recovery could help us understand how Oceania's innovations affected the evolution of the human palate.

**Australia:** It is hard to talk about the plants of Australia without mentioning the aromatic and quick-growing eucalyptus. There are a stunning seven hundred species. The sap of *Eucalyptus gunnii* was the source of the earliest Australian alcohol.

When British fleets arrived in Australia, they used peppermint gum from the *Eucalyptus piperita* tree in teas and tonics. The flavor was reminiscent of mint, and the effects were more medicinal. Today, eucalyptus notes dominate some of the boldest commercial bitters and are once again gaining steam on the international market. Other native plants in the same family as the eucalyptus, the *Myrtaceae*, include nutmeg and clove, which make delicious bitters for cocktails.

**South Pacific Islands:** The biodiversity of botanicals of the South Pacific Islands was enhanced when tectonic plates, each with their own independently evolved flora, crashed together to form the volcanic ring of fire. Today's South Pacific is a mixture of indigenous, tropical Asian, European, and global influences that contribute to the myriad of botanical infusions used for medicine, sustenance, and pleasure. An astonishingly high 20 percent of the flora of the South Pacific Islands is used by local communities for medicinal purposes.

### Europe and Western Pharmacy
In Europe, the extraction of botanical material in alcohol dates back to the wines and elixirs of the Hippocratic era in Greece. Since then, commercial bitters recipes have been passed down generations and safeguarded in vaults.

The physician Paracelsus, widely known in Europe as the father of using chemistry in medicine and toxicology, formulated an elixir that eventually inspired the development of the well-known Swedish Bitters that are still commercialized to this day. His original formula is thought to be based on three key ingredients: aloe, saffron, and myrrh. These are known to stimulate the tongue's bitter receptors and trigger the production of digestive enzymes, functions akin to a digestif.

Many of the bitter liquors that we are familiar with today—including absinthe, Campari, and Jagermeister—started out as patented medicines and digestifs. These herbal liquors are also classified as kräuterlikörs and "half bitters," because of their typically higher sugar content.

Western pharmacy was preserved and perpetuated in monasteries. Monks gathered herbs, raised them in their abbey gardens, and prepared them to treat the ill. Several

of the herbal liquors that are famous to this day were formulated by clergymen and monks, whose recipes have been closely guarded. Stoughton's Elixir, which was created in 1712 by the Reverend Richard Stoughton, was among the first medicines in England to receive a British royal patent, and it eventually became a successful export to the American colonies.

In the nineteenth century, bitters evolved to take on a new use: to support the well-being of soldiers, sailors, and colonial migrants as they explored territories, mostly in tropical climates. The classic gin and tonic was created in response to centuries of colonial expansions. The malaria-fighting property of this cocktail comes from the quinine-laden tonic water made from the bark of the South American cinchona tree.

### North America

Bitters and infusions in North America represent the juxtaposition of botanical uses by First Nations communities along with the cultural mixture and assimilation of peoples from around the world in the United States.

**First Nations:** Native Americans have used botanicals for thousands of years to heal the body and purify the spirit. Various Native American tribes have a tradition of healers carrying medicine bundles of plants used to prepare remedies for colds, fever, and pain as well as for protection against bad spirits. Some of the most commonly carried plants have been hops, American ginseng, wild black cherry, pennyroyal, sage, willow bark, dogwood, and fever wort.

**Left** The rise of bitters as patented medicines gave birth to traveling practitioners known as quacks, who traveled from town to town offering dubious remedies to treat ailments.

**North America and Cocktail Culture:**
Bitters were commonplace in American colonies and were initially imported from Europe. After the Declaration of Independence, local distillers in Boston and other cities began producing their own versions of bitters. In this era, bitters were predominantly used in the United States as medicines, but their taste and alcoholic content did not go unnoticed. The earliest definition of a "cocktail," dating from 1806, describes it as a mixed drink consisting of spirits, bitters, water, and sugar.

A century later, alcohol, food, and drug safety regulations decisively shifted the focus of bitters from medicine to recreation. The United States Pure Food and Drug Act of 1906 required companies to add the alcohol content of their products to their labels. Because that content could not be 20 percent or more, bitters companies changed tactics and began to advertise their products for alcoholic consumption. For example, Lash's Bitters was advertised in the early twentieth century to families: on a 1901 trading card, it portrays a little boy needing to use the chamber pot and suggests the use of Lash's Bitters for constipation. By 1907, advertisements for Lash's Bitters promoted it instead as a hangover cure.

The Prohibition era (1920–33) had a calamitous impact on bitters manufacturers. While bitters, along with other mixtures, were useful to improve the flavor of cheap alcohol, during this period the manufacture and sale of alcoholic bitters was illegal in the United States and almost all makers went out of business.

The past decade, however, has seen an enthusiastic bitters revival with incredible innovation in both recreational and medicinal products. Chemistry, biology, physics, and engineering come together to inspire us with innovations including spherification, cocktail gels and powders, flavor-changing cocktails, layered cocktails, and foams.

## APERITIFS AND DIGESTIFS

Aperitifs, including vermouth, sparkling wine, and pastis, are dry tasting and are drunk before meals to stimulate the appetite. In contrast, digestifs, such as Strega, are concentrated, syrupy sweet, and bitter herbal spirits which are served after meals to help with digestion.

Depending on how you use the ingredients, you can make a delicious aperitif cocktail out of both aperitifs and digestifs, as long as it's not too heavy and sweet.

These maps show where some of the major products in the bitters family come from. Many producers continue to use safeguarded recipes.

**Trinidad** •

**Angostura:** Aromatic bitters of gentian and other botanicals invented in 1824 by Dr. Johann Gottlieb Benjamin Siegert

**Brazil** •

**Brasilberg:** Bitters with a rich, fruity, and herbal flavor developed by Paul Underberg

**United States** •

**Peychaud's:** Aromatic bitters of various botanicals

**Netherlands** •

**Beerenburg:** Bitters of fifteen herbs including angelica, gentian, violet, and bay

**Hoppe:** Bitters from orange peel

**Sonnema Berenberg:** A gin-base infusion of seventy-one botanicals

**France** •

**Chartreuse:** Made from one hundred and thirty herbs and plants macerated in alcohol and steeped for about eight hours

**Suze:** Wine-base gentian bitters

• **South Africa**

**Capertif:** Chenin blanc-base aperitif from the Western Cape with Fynbos regional herbs including African artemisia, buchu, and kooigoed, along with citrus

**Ghana** •

**Alomo Bitters:** Proprietary bitters by Kasapreko

**Herb Afrik:** Gin-base bitters from a proprietary recipe

• **China**

**Shi Quan Da Bu Jiu:** Rice wine-base tonic with roots such as rehmannia and angelica, and mushrooms

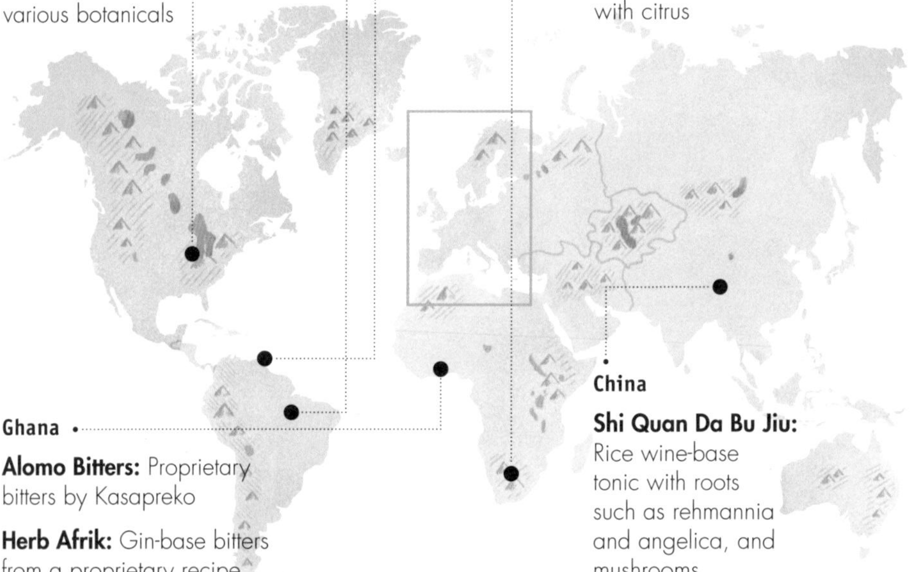

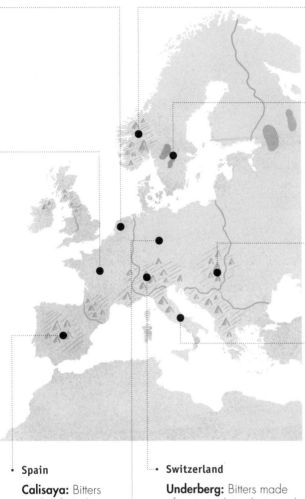

## Norway

**Marka:** Meaning "woodland forest," it contains sixteen botanicals infused and rested in oak

## Sweden

**Swedish Bitters:** Bitters and traditional herbal tonic composed of aloe, angelica root, carline thistle root, camphor, manna, myrrh, rhubarb root, saffron, senna, and zedoary root

## Hungary

**Unicum:** Herbal liquor blended with more than forty herbs and spices in aged oak casks for more than six months

## Italy

**Amari bitters:** Liqueur bitters of bark and herbs

**Braulio:** Bitters made of alpine botanicals

**Campari:** Aperitif made of sixty-eight aromatic and bitter herbs, including Chinese rhubarb, cinchona bark, bitter orange peels, and quinine

**Cardamaro:** Piemontese moscato wine infused with cardoon and thistle

**Cynar:** Contains thirteen botanicals

**Fernet Branca:** Herbal wine-base digestif bitters with thirty botanicals, including camomile, saffron, gentian, and rhubarb

## Spain

**Calisaya:** Bitters made of cinchona bark and other botanicals in a brandy base

## Switzerland

**Underberg:** Bitters made of various alpine botanicals first produced in 1846

## Germany

**Jagermeister:** Bitters made from fifty-six botanicals, including ginger, cinnamon, star anise, cardamom, and orange peel

## Chapter 2
# Bitters and the Planet

We are so crazy about plants that we have
dedicated our lives to studying, experimenting with,
and sharing information about them. Luckily, that
involves a bounty of fun with bitters and cocktails.
A living ninety-five-hundred-year-old spruce tree or the
ancient bristlecone pine forest will inspire us to collect
sap from trees of different ages to try in a cocktail.
The giant pods of the behemoth baobab tree teach us
about life in the African savanna while pairing well
with a smoky scotch. Here, we present the basics
of the botany and ecology of bitters. We all know
biodiversity is good for the planet, but a wide
range of plants also means a wide range of
flavors to play with in your kitchen.

# The Plants
# Behind the Drinks

Bitters don't just come from roots, leaves, fruit, and seeds but may derive from other parts, such as the stem, that can produce saps and barks. Different plant parts produce distinctive tastes and aromas. Different extraction methods bring out different attributes. These attributes are products of hundreds of millions of years of evolution. The resulting drink is more than just its taste and effect on us: it's a concentration of millennia of evolution. Land plants (also known as vascular plants) first appeared on the Earth around four hundred and fifty million years ago.

VANILLA PLANIFOLIA

### The Plant Kingdom

As plants evolved, they changed life on the earth. Remarkable in their self-sustaining ability to make their own food from sunlight and water, plants remain the primary source of food for most of life and provide habitats for many species. Land plants are extraordinary in that they can inhabit two distinct environments at the same time: below ground in the soil (the plant's roots) and above ground in the air and sunlight (the plant's shoots). Plants continue to evolve in their environments and must undergo rapid change if they are to adapt to increased climate variability and human activity.

### What to Call Them?

All identified plant species have a unique scientific name. Many plants also have one or more common names used by different cultures and communities though the same common name may also be used for different species.

The scientific naming is based on what is known as binomial nomenclature, composed of two Latin names. The first name identifies the genus the plant belongs to and the second name identifies the species within the genus. For example, the scientific name for tea is *Camellia sinensis*, which means it is the species *sinensis* of the genus *Camellia*.

There are eight heirarchical ranks for grouping plants: superkingdom, kingdom, phylum, class, order, family, genus, and species. Some groups are different in size and age, but they can all be traced to a common ancestor: algae. Many species have become extinct and others have radiated, evolving rapidly into multiple species.

The classification system keeps changing as new science emerges about genetic findings on how plants are related and as species that are new to science are discovered.

## The Anatomy of Shoots and Roots

To identify the correct plants for bitters, we need a working knowledge of their external and internal structure. The correct plant and plant part are most often recognizable by sight and smell, but be aware that like us, plants have specialized organs that mean different parts have different plant compounds. Tea, for instance, contains caffeine in both leaves and flowers, but only the leaves contain the sought-after flavor. (Research suggests that a high level of caffeine in the flower stamens keeps the pollinating bees healthy.)

**Plant stems** serve to transport water, nutrients, and minerals from roots to leaves, where photosynthesis occurs, and to distribute the resulting sugars from the leaves throughout the plant. Stems also have an important function in plant size and shape. Stems can grow straight up toward light or stems can grow along the ground to help the plant spread out, thus ensuring it catches the optimum amount of light.

The use of stems in bitters depends on the type and layer of the stem used. Stem layers consist of the ground, dermal (skin), and vascular (vein) tissues, and each has a different texture. The exterior dermal tissue, such as bark, provides a protective layer for the interior vascular tissue, so it is usually more tough than the interior. Examples of bark used for bitters and tonics are cinnamon, pepperbark, and willow.

**Leaves** are the site of photosynthesis, where plants convert sunlight and water into energy. Most recognizable leaves are composed of the petiole (the stalklike part that joins the leaf to the stem) and the blade (the open part that catches light). Usually, the leaf blade is the darker green side, packed with chloroplasts that are used in photosynthesis.

Not many colors have their own flavor, but green certainly does. Think about wheatgrass juice, one of the plants with the highest chlorophyll concentrations. Realizing that leaves taste green because of their chlorophyll content is important to keep in mind when extracting (infusing) leaves for bitters. You are bound to extract some chlorophyll, which can result in a green flavor profile and should be carefully balanced, depending on what taste you are going for. When bitters are stored over a length of time, the flavor of chlorophyll transforms as it oxidizes.

You don't have to be a seasoned botanist to identify the leaves of most plants. However, some leaves have been modified to be nearly unrecognizable. In cacti, for example, the stem performs the photosynthesis and the leaves have been transformed into spines.

**Flowers and fruits** are sexual reproductive organs that produce and protect seeds. Wind and water are the vehicles for pollination in some species, but animals are dominantly relied on for flower pollination and seed dispersal. These plant–animal relationships have led to the evolution of flowers and inflorescences (clusters of multiple flowers) ranging in shape from nearly invisible catkins like oaks to giant flowers like Rafflesia about three feet (90 cm) in length. Likewise, fruits range from the microscopically small *Wolffia* to the giant pumpkin.

Sometimes the whole flower is used in a bitters recipe, but other times only a specific part is used, such as the petals of rose, the stigmas of saffron, or the pollen of wild flowers. If you're making an extraction from an edible flower, it's worth tasting the different floral plants by themselves to see how each contributes to the flavor.

Fruits are formed from the flower ovaries of plants and can vary in type and structure. The classification of fruits can be deceiving because not all things that we know as fruits and berries are botanically so—juniper "berries" are actually cones, like pinecones, but with the scales merged together to protect the seed. Technically, a berry is a fleshy fruit with multiple seeds in it, such as a tomato. A drupe is a fleshy fruit with a single seed, such as an apricot. A pome is a fruit with a fleshy outer layer around the seed-containing core (a trick played by apples and quinces). The list of fruit types goes on and on, but most important is to keep an open mind that fruits may be in disguise.

**Roots** stabilize plants and absorb water, nutrients, and minerals from the soil. Because roots serve as the anchor, they are generally a strong, hardy, and harvestable material. There are two main types of roots used for making bitters. Taproots consist of a single thick root, from which a large network of smaller roots grow. Rhizomes are modified underground stems that can send out both shoots and roots. Fibrous roots, such as those at the base of orris rhizomes and coconut palm, have a dense mat of thinner roots.

## Know the Life Cycles

For bitters, it is important to understand a plant's life cycle so you will know when it is the best time for harvesting. The plant life cycle can be annual, biennial, or perennial, depending on the time it takes to produce flowers and seeds. Annual plants germinate from seed, grow, mature, bloom, produce seeds, and die within one year or growing season. A biennial goes through its life cycle in two years, where it produces leaves as well as food storage organs during the first year, and then overwinters to produce flowers, fruits, and seeds the following year. Perennial plants are those that live for longer than two years and are sometimes distinguished by woody growth.

---

### CLASSIFYING ON THE GO

Our field guide in the Venezuelan Amazonas always carries a machete on his side when walking through the forest. One of the primary uses of this knife is to help identify plants through their shoots and roots as well as to harvest them. On one occasion, intrigued by a soaring tree with reddish peeling bark, we questioned what it was. As botanists, we were used to identifying plants first by their leaf anatomy, but the leaves were too high above us, with the tree crown reaching toward the upper realms of the forest canopy. Our guide pulled out his machete and gently chipped a piece of the peeling bark. He smelled the inner moist layer and shared its name in the Piaroa language. This is how we learned that this particular tree was not the kind we were looking for. While leaves from tall trees can be obtained—we also saw our guide climb using a homemade rope harness—it is always valuable to know more than one way of identifying plants.

# Where the Flavor Comes From

Plants taste the way they do because of their evolutionary history and their mode of survival in their ecosystem. The flavor of plants comes from the phytochemicals they produce depending on the threats they encounter in their environment, from predators to extreme heat. Unlike animals, plants are rooted and must defend themselves from their stationary positions. Understanding the environments where plants grow helps you understand their properties.

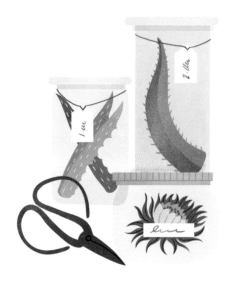

### The Taste of Change

From microscopic algae in the sea ice to succulents with flowers three feet (90 cm) wide, our world is covered with plants that sustain our ecosystems. Almost every biome (that is, a large-scale ecosystem) has its plant life; even frozen tundras contain hardy dwarf shrubs and mosses struggling to survive the harsh conditions—with many developing interesting flavors as a result.

As the climate changes, plants and people must adapt or risk peril. The warming of the Arctic is melting permafrost and making traditional livelihoods less possible; some people are turning to agriculture, looking to tundra species with potential as future cash crops. Many of these are bitter medicinals such as rhodiola.

Meanwhile, the desert biome is rapidly gaining ground. Woodland and chaparral biomes are receiving less rain, and

increasingly warm temperatures are pushing native plants uphill to cooler regions. There is opportunity to preserve certain habitats and expand desert agriculture, which may lead to fascinating new drinks as a result, such as wines from desert grapes, beer from desert dates, and California-grown agave for tequila.

### Understanding Terror

Just as plants shape the environment, environments shape plants. Knowing where a plant comes from and its ecology clue you in to its flavor character. The relationship among plants, place, and people is known as terroir and it is crucial to appreciating bitters. Microclimates, soils, and management practices influence plant attributes, particularly their phytochemicals, also known as secondary metabolites. Many of these are defense compounds.

A plant's defense system is particularly important to its flavor, because it frequently involves a chemical armor of secondary metabolites. These metabolites cause plants to taste foul and even be poisonous. Humans often enjoy the flavors, but we need to pay attention and understand which plant compounds are toxic to us and at what levels.

Ultraviolet (UV) light from the sun, heat, fire, drought, wind, and rain all affect the chemistry of a plant, as well as its appearance and texture. Sun stress can make leaves purple, drought can cause some plants to flower early, and too much rain can make fruits bland. Physical stress and tissue damage often make a plant release hormones, such as jasmonic acid, which turns on defense pathways. The same jasmonic acid helps grapes improve their aroma for better wine or adds floral attributes to pisco. People can exploit this process by physically beating up the plants.

There is also the more invisible side to terroir: the small critters, microbes, and minerals in the soil. These microbes not only interact with the roots but cover the leaves, stems, and flowers. Many of the microbes on plants are the fungi used in the fermentation process. Alcohols from mezcal to wine can be made using locally occurring yeast.

Finally, climate change affects the flavor of plants by altering the conditions and community of the plant. Climate change also alters where plants can thrive in the wild and under cultivation. Scientists, farmers, and food sellers have only just begun to track this relationship and there is much to learn.

**Below** Plants growing in the Himalayas have been found to have higher amounts of phytochemicals compared to the same species found at lower elevations because of the heightened stress at higher elevations.

## Plant Domestication

The majority of the plants that humans have used for food, beverages, tonics, and other medicine come from the wild and represent our roots as hunters and gatherers. Starting ten thousand to twenty thousand years ago, humans began to intentionally transplant seeds and whole plants from their original settings, beginning the dynamic process of domestication. Two defining aspects of domestication are that humans selected for plant qualities they preferred, and plants had to change to thrive in a new environment. Some of the earliest plants to be domesticated were fig trees, wheat, millet, flax, peas, barley, and gourds.

Domestic plants changed from their wild ancestors in many new ways, particularly in their phytochemicals and associated flavor profiles. While there are hundreds of thousands of domesticated varieties of food crops, when it comes to bitters we shouldn't limit ourselves to cultivated food plants; there are thousands of species found in the wild and used in horticulture.

## Diversity in Our Drinks

Biodiversity is the variation among living organisms and their environment. Once it is lost or forgotten, it is hard to bring back. At present, we use only a small fraction of those plants that have been historically used for food, beverages, and medicine. Of the more than three hundred thousand species of plants identified on the planet from six hundred and twenty botanical families, more than seventy-five thousand have been used for food and medicine at some point in time. The three most prevalent botanical families are the *Asteraceae* (sunflowers), *Fabaceae* (legumes), and *Orchidaceae* (orchids); together they make up around 25 percent of all flowering plants. It should therefore be no surprise that these families are well represented in bitters

and cocktails. It is estimated that we commonly use only three hundred species in twenty botanical families for food. Preparing bitters from underutilized plants is one way of introducing biodiversity back into our diets. This isn't just good for the planet and our health, but for our pleasure, too. Biodiversity lends bitters an incredible range of flavor. For example, at the genetic level, tea has two main commercial varieties: the broad-leaved assamica variety, which evolved in subtropical regions to be more bitter, and the sinensis variety, which evolved in more temperate regions for sweeter taste notes. Within each variety, however, there are hundreds of more locally developed varietals that have their own enjoyable flavors.

Human activities that are not sustainable, such as overharvesting wild species, have reduced the abundance of many species that were once common. Many consider that we're experiencing the sixth mass extinction of biodiversity on the planet. However, human activities can also aid the conservation and even creation of biodiversity. For instance, some farmers can grow more varieties and species of plants while promoting wild diversity around their farms. We strive to use ethically sourced ingredients ourselves and believe conscious consumers can help support conservation of our planet's species.

## Sourcing Sustainable Botanicals

There is a huge range of ingredients that you can find in your local grocery store or safely forage from your yard or in the wild. However, as you become more experienced and expand your use of botanicals, we suggest taking the following points into consideration:

01. **Direct and Knowledgeable Sourcing.**
    Learn as much about the sourcing of botanicals as possible. We encourage you to build relationships where you

source botanicals, as well as to form a community with others who source diverse botanicals to share experiences of product quality.

02. **Conservation Status.** Make sure the botanicals you use do not have a vulnerable status. Refer to the IUCN Red List of Threatened Species (available online) to double check the conservation status of a botanical.

03. **Sustainable Agricultural Practices.** Try to source cultivated botanicals from farms that use sustainable agricultural practices, such as organic, biodynamic, and diversified farming.

**Above** Some cultivation practices, such as organic coffee farming in Peru (pictured here), contribute to sustainability through their lack of agrochemical use while also promoting biodiversity by growing tree-based cropping along farm edges.

04. **Equitable Benefits to Harvesters and Farmers.** Try to source botanicals in ways that support harvesters, farmers, and communities.

05. **Always Ask.** When sourcing botanicals, always ask, "Where does this come from? Who harvested it? Under what conditions was it harvested?"

For more details on how to find ingredients, turn to page 167.

# Botanical Safety

Botanicals have been used in food, drink, and medicine for many thousands of years. If something has a long history |of use and reports of low toxicity, you can usually consider it safe if used in the same mode and dosage.

Any substance, even water, can become toxic in excess. Freshly ground nutmeg, the seed of *Myristica fragrans*, contains myristicin, which can have potentially lethal interactions with other food and is dangerous in large quantities. A few dashes over the foam of a latte create a pleasant holiday spice, but two or three spoonfuls of nutmeg can be fatal.

Botanical safety varies from person to person. Many botanicals are dangerous for pregnant women, for instance, and some are harmful if mixed with certain prescription medications.

Sometimes, harmless plants can look a lot like their poisonous relatives, and mix-ups can end in death; in fact, poisoning from botanicals is often due to misidentification during harvest or sale.

Still other botanicals can pose safety threats based on their method of preparation; some

toxic compounds, for example, can be pulled out by using only certain extraction methods. If records of a plant's use indicate it is prepared as a water extract, such as a tea or tisane, it is safest to follow the original extraction method over extracting with alcohol.

Even with widely used botanicals there have been cases of toxicity. Sometimes, the danger is not related to the plant itself but to where it is growing. Plants can accumulate heavy metals from contaminated soils or can contain pesticide residues. We encourage people to buy from sources that you trust.

Sometimes, what you find in the market are adulterated botanical products, which means plant species were substituted or added to the product without being advertised to reduce cost. With the broader availability of material from increasingly untraceable Internet sources, it is something to be wary of; a trusted supplier is always best.

You can increase your knowledge on plant safety through resources developed by botanical and herbal councils as well as national agencies. International organizations and industry associations have also developed their own versions of botanical safety handbooks that brings together much of this information. For example, the United Nation's World Health Organization (WHO) launched the Good Sourcing Practices for supporting safe botanical material in a way that is also sustainable for people and the environment. The more you can inform yourself, the better off you will be.

There is a lot of misinformation in online sources and not even published studies are necessarily the final say on any matter. Plant knowledge is ever growing. When looking for new plants or plant parts to concoct bitters, go about your discoveries with an inquisitive, curious, and critical mind. Try to find multiple reputable sources that support a claim about a plant.

We do not recommend consumption of bitters to anyone who is pregnant, who has moderate-to-severe medicinal conditions, or is taking medications.

While bitters are fun, we must always proceed with safety in mind.

## LOOKING BEYOND THE LABEL

A grocery store study by Dr. Mark Stoeckle used DNA barcoding to test herbal tisane products. The result: 30 percent of them contained botanicals that were not listed. In contrast, a study on loose herbs from specialty shops showed the large majority of products were correctly labeled. The direct relationship store managers have with purveyors and farmers can lessen the risk of adulterated or incorrect material, so look for businesses that care about their communities.

# Bitters and the Body

Bitters have been used since prehistoric times, and bitter infusions of medicinal plants are still widely used to treat and prevent illness in health-care systems around the world. Today, approximately seven thousand modern medicines are derived from bitter plant medicines. Bitters can be classified on the basis of their physiological effects on the human body, with digestive bitters being among the largest class.

### Medicinal Bitters

The bitterness of bitters mostly comes from some well-known phytochemical classes: alkaloids, phenols, polyphenols, flavonoids, isoflavones, terpenes, and glucosinolates. All of these are also known for health benefits, including having antifungal, antiseptic, antidepressant, cardio-protective, hormone regulating, immune boosting, and blood sugar-regulating properties. Among the most powerful effects for preventative health are antioxidant and anti-inflammatory activity; because inflammation can lead to chronic diseases and many types of cancer, it's not surprising both botanists and doctors are looking to bitters.

Less than 20 percent of commonly used medicinal plants have been comprehensively researched for their efficacy, and even among those that have been researched, it is another story when they come to be mixed in bitters. A bitters made of whole herbs of caraway,

fennel, and anise, for instance, is known to have antispasmodic activities, but the isolated essential oils of each of the plants do not have this function and can even induce an opposite effect.

### Digestive Bitters

Bitter tasting compounds often are helpful to the digestive system. The bitter reflex can be thought of as a series of stimulation and secretion throughout the body. The appetite and certain mechanisms repairing the lining of the intestines are stimulated, and enzymes and bile that aid detoxification, plus pancreatic hormones that regulate blood sugar, are secreted.

This explains why digestive bitters are one of the largest classes of bitters in traditional medicinal systems, known to help with constipation, gas, bloating, loose stools, food allergies, and acid reflux. In addition, the bitter reflex is considered to improve nutrient and mineral absorption, increase the appetite, promote healthy blood sugar levels, protect liver function, and heal inflammatory damage to the intestinal walls.

Digestive bitters can be taken before or after a meal, depending on their function and flavor. Many digestive bitters are taken before a meal to prepare the body for eating; other digestive bitters with strong aromatic profiles are taken after a meal to help digestion as well as to freshen the breath.

Bitters, herbs, and other therapeutic products used to relieve bloating and intestinal gas are known as carminatives. Many carminatives have rich volatile essential oils; for example, peppermint oil has been shown to serve a carminative role to relieve bloating. Likewise, the peel of bitter orange is frequently found in bitters not only as a flavoring agent but also as a mild carminative.

Several species in the celery family, *Apiaceae*, have seeds with known carminative effects that are often taken following meals. These include fennel, anise, dill, caraway, and cardamom; "Mukhwas" in India and amaros in Italy are popular after-dinner examples.

### Bitters for Stress

There is a long history of botanicals being used for their anti-anxiety and relaxant properties. Exactly how they work varies from compound to compound, but broadly speaking, they tend to work one a few ways.

*Adaptogenic* bitters boost your body's ability to adapt and recover, improving your general well-being. Some common adaptogenic ingredients are ashwaganda, American ginseng, and Siberian ginseng.

Other bitters have a more targeted effect. *Anxiolytics*, for instance, are mild sedatives that relieve apprehension without affecting the body's faculties as a whole. Sedative plants in general can function as *analgesics* (relieving symptoms of pain); common examples are often found in bitter tonics include hops, valerian, wormwood, chamomile, and mugwort. There are also *nervine remedies*, which means the plants have a calming effect on nerves and stress.

### PLAYING SAFE

If you are on any medication, be aware: bitters can interact with or affect the absorption of medicines. Medicinal plants used in bitters can also have adverse effects if used incorrectly, such as through overdose or improper preparation, or else because you unknowingly bought contaminated or adulterated material. Research and source carefully, and if you are not sure, seek medical advice and err on the side of caution.

Chapter 3
# A Matter of Taste

A gentian extract we served at a bitters-making workshops had some participants cringing—but others wanted to sprinkle it on everything. While most recognized the extract as tasting bitter, the level of bitterness perceived, as well as the preference for this taste, varied from one person to the next. It has been remarkable to learn what people like or dislike, and how this relates to their genetic and cultural backgrounds. Our perception of flavor is a complex sensory experience. In this chapter, we'll discuss just how those senses work.

# Basic Taste Sensations

Flavor is a common experience that can unite us all. To create delicious and intriguing bitters and cocktails, it is useful to understand the principles of flavor. The five basic taste senses are widely recognized as sweet, salty, sour, bitter, and umami. The flavor we perceive is a combination of what we taste and what we smell, but we harmonize these sensations into a single experience.

Of course, no one could describe all the diverse flavors of bitters using combinations of only these five basic tastes. Flavor isn't just taste: it's also aroma, along with the influence of touch, sight, and sound. In many cases, aroma is present in trace amounts, yet it is the defining characteristic of a particular drink. The unique experience of bitters comes from the interaction of the basic taste senses with hundreds of aromatic compounds, such as those that have floral, citrus, spicy, smoky, and other notes. While we "smell" to perceive aroma and "taste" to detect the five basic tastes, we simply don't have a verb for the perception of flavor as a combination of taste and smell.

## Why Do We Have Taste?

Taste is an ancient adaptation. The earliest vertebrates from which we evolved arose in the ocean more than five hundred million years ago, and they already possessed the ability to taste. As different vertebrates diverged in their evolution, all maintain some form of taste receptors.

Bitterness is thought to signal whether foods are toxic or medicinal. Healing systems around the world have identified toxic and medicinal plants through perceiving bitterness. Sometimes a plant species can be both, and there is a narrow threshold between a dangerous and a therapeutic dose.

Sweetness, on the other hand, indicates the availability of energy and calories essential for survival. Sourness serves as a detector of acidity that can signal either food spoilage or nutritious foods, such as fermented foods. Umami is linked with survival by enabling us to identify dietary amino acids that our bodies need for building proteins.

The science around saltiness is unresolved. A possible role is to help us moderate sodium intake and regulate electrolytes.

## The Basic Tastes

For centuries, people have puzzled over the science of flavor and what basic tastes we can all perceive.

According to traditional Chinese medicine (TCM), the five tastes are bitter, salty, sweet, sour, and pungent. Each of these tastes is considered to nourish a specific organ or organ system in the human body and is associated with seasons and warming or cooling energy.

In the Western world, four tastes were historically recognized; bitter, sour, salty, and sweet. The ancient Greek philosopher Democritus (460–ca. 370 BCE) theorized that the sense of taste was related to the shape of food particles. Sweet atoms were thought to be large and spherical, sour atoms were largish polygons with a rough texture, salty atoms had the shape of an isosceles triangle, and bitter atoms were small spherelike polygons with a smooth texture.

Plato (late 420s–late 340s BCE) built on Democritus's flavor work. He hypothesized

that our ability to taste was based on taste atoms of foods and beverages that penetrated the capillary vessels in our tongues, and that these vessels were connected to the heart, which determined our preferences for flavor. Plato's taste theory is one of the earliest recorded key-and-lock models of taste, where the capillaries were viewed as keyholes on the tongue that taste compounds of a specific shape could enter.

Aristotle (384–322 BCE) proposed four basic tastes in his treatise *De Anima* (On the Soul): sweet, sour, salty, and bitter. These four tastes remained largely undisputed by scientists for two thousand years until the recognition of the savory taste of umami by the Japanese chemist Kikunae Ikeda (1864–1936).

In 1908, after being intrigued by the delicious taste of the dashi broth that his wife prepared, Professor Ikeda isolated the crystals of the amino acid, glutamic acid, which provided the umami flavor, and developed monosodium glutamate as the chemical basis of the fifth taste. He called this fifth taste *ajinomoto*, or the "essence of flavor." Today, *umami*, meaning "delicious taste" in Japanese, characterizes the savory taste of meat, cheese, mushrooms, and some vegetables. Although Professor Ikeda declared umami as a fifth taste in the early twentieth century, it was only in 1985 that umami was accepted in Western science.

In addition to the five widely recognized basic taste senses that we perceive in our mouths, scientists have more recently proposed additional tastes, including calcium, fat, carbon dioxide, metallic, and *kokumi*. Kokumi is a mouthfeel perceived by calcium channels on the tongue that enhances the flavors sweet, salty, and umami and derives from the Japanese words for "rich" and "taste."

# Taste Plus Aroma

We experience flavor through two systems: the olfactory system, which processes scent molecules through our nasal cavity, and the gustatory system, which contains taste receptors in our mouths. From these sensations, our brain constructs a complex flavor. There is great genetic diversity in our taste receptors and even greater diversity in our olfactory receptors, creating near infinite possibilities in the construction of flavor.

## Taste Receptors

In the iconic tongue chart, the tongue is divided symmetrically and mapped into color-coded hemispheres. The back of the tongue is mapped out with the ability to taste bitter, the tip of the tongue is mapped with the ability to sense sweet, and the sides of the tongue are mapped with receptors to discern sour and salt. Depending on when it was designed, it probably does not include umami. Current science has debunked the concept of the tongue map, favoring a concept of distributed taste receptors on the tongue.

Taste receptor cells (also known as gustatory cells) are located in taste buds on the upper surface of the tongue, cheeks, soft palate, upper esophagus, and epiglottis. Most humans have two thousand to eight thousand taste buds. Most taste buds sit on raised mushroomlike protrusions on the surface of the tongue, known as fungiform papillae. These appear like little pink dots.

The most numerous papillae on the tongue, however, are filiform papillae, but these do not contain taste buds. These fingerlike structures work to spread a solution of food, beverage, and saliva across the papillae that do have taste buds. They help what we eat or drink to come into contact with our taste receptors, which trigger the sense of taste; anything that can activate them is known as a tastant. The signals these receptors send to the brain are then processed in the gustatory cortex.

Scientists have identified two families of taste receptors: type one receptors for the sense of sweet and type two receptors for the sense of bitter. While there is only a single human sweet receptor, more than twenty human bitter taste receptors have been identified, and these let us perceive a greater diversity of bitter compounds compared to other tastes.

## Processing Scent

The olfactory system is the most primordial method our bodies have for reacting to our environment. It is also one of the most crucial; for instance, without a sense of smell we would have difficulty telling safe food from rotten. The olfactory system is one of the most developed human senses connected to memory, emotions, and survival.

Our ability to smell aromatic phytochemicals in bitters is largely due to the olfactory epithelium, a section of specialized tissue approximately the size of a small coin at the top of the nasal cavity. However, although the human olfactory epithelium is comprised of hundreds of different olfactory cells, the human brain allocates only 0.1 percent of its space to the sense of smell. In contrast, the human brain dedicates up to 25 percent of its space for the sense of sight. When it comes to experiencing flavor, more complex factors are at play.

## CAN YOU TASTE WITHOUT SMELLING?

This activity helps you perceive how flavor is comprised of both taste and aroma.

01. Find a piece of mint leaf or other aromatic food or beverage, such as aromatic bitters or flavored candy.

02. Take a deep breath and then briefly pinch your nose as you prepare for your tasting experience.

03. Keeping your nose pinched, place the piece of mint leaf or other selected aromatic food in your mouth and chew for 5 seconds. What taste and aroma do you perceive? Is it characteristic of mint? You can spit out the mint leaf after this step or swallow it.

04. Take another deep breath, this time keeping your nose unpinched. Place another mint leaf (or your selected aromatic food) in your mouth and chew for 5 seconds. What taste and aroma do you perceive? Is it characteristic of mint? You can spit out the mint leaf after this step or swallow it.

05. Compare the two flavor experiences of tasting the mint leaf or other selected aromatic food with your nose pinched and unpinched. This exercise should demonstrate how crucial scent is to your overall flavor experience.

This activity can be carried out with a group of family and friends by giving them different-flavored candies and having them guess the flavor with their eyes closed during the tasting.

## Other Influences

The holistic perception of flavor includes the crunchy noise of crackers, the creaminess of butter, the coolness of ice cream, and the bubbliness of tonic water. It can also include attributes of the environment, such as the visual appeal of a cocktail or the space in which we are enjoying it. It is important to keep in mind that our flavor experience is also determined by these other sensations.

In addition to actual temperature, some compounds in plants can modify our perceptions of the temperature of foods and beverages. The "hot" sensation of spices is known as piquance; the key receptor for this is found all over the body and sends a heat signal to the brain even if, say, the chili pepper you are eating is actually cold. In contrast, menthol from a fresh mint cocktail results in a cooling sensation whatever the real temperature of the drink.

## You Are What You Taste?

What is your favorite basic taste? It should come as no surprise that we at Shoots & Roots prefer the taste of bitter. What tastes pleasant to us, however, can taste unpleasant to you. Furthermore, what you perceive as unpleasant now can taste pleasant to you when you are hungry.

Our ability to perceive tastes is anatomical, but our reaction to what the flavors we perceive is relative and varies based on genetic and socioecological factors, such as our cultural heritage and the environments in which we were raised. Genetics, for example, can affect whether you can taste the common bitter compound phenylthiocarbamide (PTC): approximately 30 percent of Caucasians can perceive it, whereas studies show that native populations of parts of Asia, South America, and Africa are much more likely to identify PTC.

In addition, many of our taste preferences are learned. Fetuses in the womb and breastfed babies can taste what their mothers eat. Other preferences are cultural. For example, vanilla is served as a sweet food in the West although it doesn't have sweet-tasting compounds; in east Asia it is used in savory dishes.

## Healthy Choices

In some ways our current food environments are providing more sources of flavor than those of our ancestors. However, many of these new flavors are coming from the processed food industry while diverse flavors from plant biodiversity have been dramatically reduced. The food industry has responded to the human evolutionary preference and cravings for sweet, salt, and fat by loading highly processed foods and other items with compounds that provide these sensations.

For better or worse, our flavor preferences and food choices have notable public health implications. The increased consumption of highly processed foods is directly linked to the increase of diet-related chronic disease, including type-two diabetes, cancers, and cardiovascular disease.

In our search for incorporating diverse, healthy, edible, and healing plants into our diets, we can draw on examples from those subsistence communities living in rural, forest, or mountainous areas that continue to rely on their surroundings for well-being. Many of the bitter plants cultivated locally are perceived by modern cuisines as unpalatable, but many human societies globally have discovered that these seemingly distasteful properties are the same attributes that support human health.

The greater the incorporation of the five basic tastes in cuisine, the greater the sensory pleasure received from these meals and the faster people feel satisfied without overeating. We think Thai cuisine is a great example of incorporating the five tastes. Because the basic tastes arise from different nutrients and phytochemicals, the meals will also probably be more nourishing.

## SPOT THE DIFFERENCE

This activity shows how taste is influenced by other sensations, including temperature and carbonation. It requires tonic water or another sweetened carbonated beverage.

*Effect of temperature on the sweetness of tonic water:*

01. Refrigerate one sealed bottle of tonic water until it is cool and keep one at room temperature.

02. Pour a small glass from each and sip, and then compare the flavor experience from the two bottles. Which one is sweeter?

*Effect of carbonation on the sweetness of tonic water:*

03. In the next part of this exercise, compare a sealed bottle of bubbly tonic water with an opened bottle of tonic water in which the carbonation is flat.

04. Pour a small glass from each, sip, and compare the flavor experience from the two bottles. Which one is sweeter?

Although the tonic water has the same amount of sweet compounds in all parts of this exercise, you should perceive that it tastes more or less sweet, depending on the temperature and carbonation. The warmer temperature should enhance the perception of sweet, as should the lack of carbonation.

# The Amazing Complexity of Bitters

In our workshops, we've tasted things you'd hardly believe could come in a glass, from barnyard and leathery to baked fruit, creosote, and so much more. The explanation? The phytochemical profiles of the plants. While plants produce a plethora of phytochemicals, we perceive primarily those that are volatile and aromatic as contributing to scent and ultimately flavor.

### The Effect of Mixtures

Before discussing flavors, it's worth pointing out that how a compound tastes on its own is not the complete picture. Combining different phytochemicals in bitters can have what are known as masking, synergistic, additive, or antagonistic effects. An example of the latter is known as "mixture suppression," in which two or more tastes can seem much less intense when mixed together than if they were tried separately. Sometimes the interaction is only in one direction. For example, salt has an antagonistic effect on bitter, but the addition of bitter has no effect on perception of salt. Bitter can also have an additive effect, for instance in enhancing perception of sour. Sometimes the effect is mutually suppressive: high concentrations of bitter and sweet cancel each other out. Couple this interaction of compounds with the variability between people's thresholds for perceiving different tastes, and you can begin to appreciate the complexity of interactions in a drink.

### The Taste of Compounds

More than six hundred bitter compounds that humans can perceive have been identified, both from well-known compounds found in common ingredients, such as tea and cocoa, and also in hundreds of less familiar species. In addition, more than five hundred aromatic volatile compounds have been identified that produce our range of sensory experiences in bitters and other herbal extracts. Many of these aroma compounds can be broadly classified across categories such as fruity, floral, herbal, spicy, vegetal, nutty, mineral, and sweet.

How you categorize a bitters aroma is yours to judge. The chart on the following pages gives you a start. It covers the common compounds found in bitters, their associated flavors, and plants that contain these. This chart will be a useful reference point when you turn to the recipes in Chapter 4.

# Common Bitter Compounds

| Phytochemical | Flavor | Example of plant sources |
|---|---|---|
| **Alpha-pinene** | Pine, coniferous | Fir, pine, chamomile |
| **Alpha-terpinyl acetate** | Herbaceous, sweet and refreshing odor, spicy, bergamot and lavender aroma, piney notes | Sweet orange, cardamom |
| **Alpha-thujene** | Woody | Frankincense, eucalyptus |
| **Anethole** | Anise | Sweet fennel, anise, star anise |
| **Beta-myrcene** | Spicy, woody | Bay laurel, hops, verbena |
| **Beta-phellandrene** | Spicy | Ginger, eucalyptus, mint, lavender |
| **Beta-pinene** | Dry woody aroma, turpentine odor, piney | Allspice, sweet orange, lemon, chamomile, lime, ginger, nutmeg, mace, fennel, rosemary, sage |
| **Bornyl acetate** | Green, floral, herb, pine | *Salvia fruticosa*, carrot, rosemary, sage, lavender |
| **Camphene** | Pungent aroma | Ginger, tea |
| **Caryophyllene** | Peppery, spicy | Cloves, hops, tea |
| **Cinnamyl alcohol** | Sweet, spicy, green, hyacinth, cinnamic | Balsam pine, cinnamon, tea |
| **Decanal** | Floral, fried, orange peel, tallow | Coriander, cilantro, buckwheat |

| Phytochemical | Flavor | Example of plant sources |
|---|---|---|
| **Gamma-terpinene** | Bitter, citrus | Coriander, lemon |
| **Geraniol** | Floral, geranium, roselike scent | Palmarosa, coriander, lemon, rose, tea |
| **Geranyl acetate** | Sweet, fruity-floral, rosy, green, lavender-like odor | Carrot, coriander |
| **N-hexanal** | Fresh, grassy, fruity, fatty, leafy, sweaty, apple, woody nuance | Tea, Chinese violet, stevia |
| **Limonene** | Lemon aroma, citrus | Sweet orange, cardamom, gentian, caraway, cinchona, coriander, lemon, bitter orange |
| **Linalool** | Sweet, floral with a touch of spiciness | Bergamot, French lavender, cardamom, gentian, tea, coriander, chamomile, mint, cinnamon |
| **Myrcene** | Pleasant aroma, lavender notes | Bitter orange blossom, coriander, lemon, chamomile, tea, hops |
| **Nerol** | Fresh sweet rose aroma (fresher than geraniol) | Sweet orange, cannabis, chamomile, tea, lemon grass |
| **Nerolidol** | Fruity, woody, fresh bark scent | Tea tree, ginger, tea, lavender, jasmine |
| **Octanal** | Fruitlike odor, pungent aroma, fatty taste | Truffle mushroom |
| **Sabinene** | Peppery, spicy | Mace, cubeb pepper, cardamom, black pepper, carrot |
| **Terpinolene** | Sweet, pine, herbal, anisic, lime, fresh, piney citrus | Pine |

# Enjoying Your Drink

Bitters have a widespread and ancient use as a way to extract and administer the medicinal properties of plants. At other times in history, such as the Prohibition era, they became a convenient guise for administering doses of alcohol through the ruse of medicine. Commercial bitters are increasingly recognized as important components of cocktails, shrubs, and sodas that can elevate drinks into an exquisite sensory journey.

### How to Taste Bitters and Cocktails

There is no right or wrong way to enjoy a beverage. However, to take in as much sensory detail as possible, you can follow certain processes to compare and evaluate your own creations. Similar to wine, tasting bitters and cocktails has four stages.

Tasting starts with sight: Is the liquid transparent or translucent, dark or light? What is the color? Does the infusion look clear or does it have visible particles?

Next, test the aroma by wafting the bottle or glass in front of your nose. Inhale and exhale deeply. While there are hundreds of aroma compounds, it is helpful to think of them according to broad aroma categories, such as fruity, floral, spicy, nutty, or sweet. However, never limit your flavor experience to these given terms; let your imagination best describe what it perceives.

Now it's time to taste the bitters or cocktail. Some concentrated bitters should be tasted a drop at a time, while others, just like cocktails, should be slurped. Swish the beverage around your mouth on all sides of the tongue and inner cheeks. Pay attention to the taste sensations as well as what it feels like (including structure and texture).

Next, inhale deeply through your mouth, then exhale through your nose with your mouth closed. This is a retronasal breath that gives a "mouth smell" and lets you better discern the aroma at the back of the throat. After your first taste and initial reflection on flavor, take another sip and deeply inhale and exhale; bitters and cocktails can transform after their first contact in your mouth. Once consumed, note the sensations of the aftertaste, also known as the finish. Finally, record notes describing the taste, aroma, intensity, aftertaste, and any other observations from your sensory experience.

Do you ever feel like the first sip or taste is the most intense? There is science to support that hunch. With time and exposure, several things are happening in the drink and in your mouth.

We experience a phenomenon called "taste adaptation," where the perception of a taste or smell gradually declines with exposure. Meanwhile, in your glass, the cocktail is transforming as well. As the bitters mix into the drink, the effects of solubility and temperature from the melting ice can change the physical chemistry of many volatile compounds, transforming their qualities as well as their potency and diluting the flavors of the cocktail. The cooling effect of the drink on your tongue also changes and mutes your perceptions of flavor.

Pay close attention to your initial impressions of the drink, closest to the point in time when everything has been muddled, mixed, shaken, and served.

## COMBINING TASTES

This activity demonstrates interactions between different flavor compounds using a bottle of readily available highly bitter-tasting bitters and a few pinches of salt and sugar. Rate the intensity of taste on a scale from one to five, with one being the least intense taste.

01. Place a drop of bitters on the center of your tongue and swish it around with saliva for 5 seconds. Record the intensity of taste that you perceive after 5 seconds on a scale of one to five.

02. Place a small pinch of salt on the center of your tongue and swish it around with saliva for 5 seconds. Record the intensity of taste that you perceive after 5 seconds on a scale of one to five.

03. Place another drop of bitters on one side of your tongue and a small pinch of salt on the other side. Close your mouth for 5 seconds while you perceive the taste and intensity of the two sensations. Record the intensity of each.

04. Now place a drop of bitters and a pinch of salt together on the center of your tongue. Be sure to use the same quantities as before. Record the intensity of each taste after 5 seconds on the scale of one to five.

05. What do you notice about the quality and intensity of the bitterness and saltiness during each tasting experience? Did it matter where the bitters and salt were placed on your tongue?

06. Eat an unsalted cracker to clear your palate, followed by drinking some water.

Repeat steps one to five using sugar instead of salt. How is the intensity of bitterness different depending on whether you add salt or sugar?

Chapter 4
# How to Make Bitters

We learned how to make bitters from multiple communities where we have carried out field research, and nod to the elders that took us under their wings. Back in New York, we supplemented what we were taught in communities with what we learned in our labs and scientific literature. We began developing our own methods to make bitters at home. In this chapter, we will teach you how to do the same, sharing both the adapted bitters recipes from our company and those inspired by ancient recipes and classic concoctions.

# Concocting Bitters

Are you ready to create some bitters?
To start you off, there are some easy
recipes using botanicals that you can
infuse directly into vodka, filter, and
drink without much processing. Once
you're comfortable with the basics, you
should be ready to start sourcing and
using more exotic ingredients to create
bitters that are inspired by plants from
around the globe. Finally, you'll find our
universal method, which can be used
to make our signature recipes and our
takes on classic bitters.

## Equipment

While you can get fancy by using a sonicator
to extract botanicals (see page 165), you
actually don't need much equipment: a mortar
and pestle or a spice grinder; a French
press or nut milk bag; some wide-mouth jars
and bottles of different sizes with air-tight
lids; and glass bottles with glass droppers
(1–4 fluid ounces/30–120 mL) to store the
final bitters, are all you need if making
traditional concentrated bitters.

Because the quantities of botanicals used
can be extremely small, we recommend that
you buy a kitchen weighing scale that is
precise to 1 gram. Kitchen scales that provide
measurements in increments down to 1 gram
(just a little more than $1/32$ ounce) are now
available online and in the kitchen section of
some major department stores. You can also
use a hydrometer to measure the alcohol
content (ABV) in your bitters.

## Converting to Grams to Ounces

If you already have a kitchen scale but
it only measures in ounces, you'll need to
convert from grams to ounces. You can
do this by multiplying the grams by 0.035.
However, keep in mind that kitchen scales
that weigh in ounces usually are accurate
by only $1/8$-ounce increments.

## Sourcing Ingredients

At Shoots & Roots, we want to encourage
biodiversity through the use of a wide variety
of ingredients from around the world instead
of relying on a limited selection of heavily
cultivated crops. For this reason, some of the
ingredients won't be found in major grocery
stores and you may have to visit specialty
stores that sell ingredients from a particular
region or cater to a specific ethnic group.
If you are unable to find them in your home
town, there are numerous online retailers that
can supply dried ingredients. See page 168
for a selection of suppliers.

# Beginner's Bitters

These recipes can be made with materials that are relatively accessible, and with basic kitchen equipment. They are extremely versatile to use in cocktails, sodas, shrubs, and foods. Each recipe yields 200 mL (⅞ cup).

## Xocolatl Bitters

This is our homage to the world's earliest known bitter cocoa beverage. *Xocolatl*, he origin of chocolate, means "bitter water," but was a drink shared by kings and Mayan Gods. Note: almond is a replacement for other seeds in the ancient recipes that are psychoactive.

4 g black peppercorns

2 g whole or ground annatto seeds

7 g almond extract (use bitter almond extract if you can get it)

30 g cacao powder (unsweetened)

2 g chili powder

⅞ cup (205 mL) 40% ABV grain alcohol, such as vodka

**Method:** Coarsely grind the peppercorns and annotto seeds (if using whole) with a mortar and pestle. Transfer the ingredients along with the cacao and chili to an 8-oz (240-mL) jar, add the alcohol, and seal the jar with the lid. To infuse the ingredients, first shake the jar, then leave it to sit for 1 week, shaking and rotating it daily.

Strain the mixture through a nut milk bag or French press and discard the solids. Bottle the bitters.

## Turkish Coffee Bitters

Traditional Turkish coffee is a potent and flavorful drink. Typically the beans are finely ground and the coffee is served unfiltered, with a thick sediment at the bottom of the cup. Cardamom is often used as a flavoring. Here we add chicory to round out the coffee notes and add a more interesting bitter flavor profile.

10 g dried chicory root or 40 g of chicory coffee substitute

5 green cardamom pods, smashed

1 star anise

30 g coarsely ground Arabica coffee beans

⅞ cup (205 mL) 40% ABV grain alcohol, such as vodka

30 g granulated sugar (optional)

**Method:** Put the chicory, cardamom, star anise, and coffee beans into an 8-oz (240-mL) jar, add the alcohol, and seal the jar with the lid. To infuse the ingredients, first shake the jar, then leave it to sit for 1 week, shaking and rotating it daily.

Strain the mixture through a nut milk bag or French press and discard the solids. If a sweeter bitters is desired, add the sugar, shake, and let sit to dissolve. Bottle the bitters. Try it in savory dishes like in rubs and yogurt, or with liquors like arak or ouzo.

## Grapefruit Peel Bitters

Grapefruit bitters are versatile and bring a balance of bitterness and aroma. This bitters contain numbing Sichuan peppercorns that bring out the nootkatone compound in grapefruit.

50 g fresh grapefruit peel

10 g dried juniper berries

3 clove buds

3 dried allspice berries

4 green Sichuan peppercorns

6 fresh mint leaves

⅞ cup (205 mL) 40% ABV grain alcohol, such as vodka

30 g granulated sugar (optional)

**Method:** If you are using commercial fruit, first rinse with a mix of lemon juice and baking soda, then scrub with a brush to remove any waxes. Remove the peel with a paring knife, avoiding the white pith. Put the grapefruit peel, juniper berries, cloves, allspice, peppercorns, and mint leaves into an 8-oz (240-mL) jar, add the alcohol, and seal the jar with the lid. To infuse the ingredients, first shake the jar, then leave the jar to sit for 1 week, shaking and rotating it daily.

Strain the mixture through a nut milk bag or French press and discard the solids. If a sweeter bitters is desired, add the sugar, shake, and let sit to dissolve. Bottle the bitters. Use creatively.

Juniper

## Celery Bitters

This blend draws from four species in the Apiaceae family: coriander, cumin, celery, and fennel. Together, these plants yield a strong vegetal, spicy pungency that works well with gin or in accompaniment with a savory cocktail.

7 g coriander seeds

15 g celery seeds

10 g dried cut lemongrass

1 g ground nutmeg

1 g ground cumin

1 pinch sea salt

50 g fresh fennel bulb

⅞ cup (205 mL) 40% ABV grain alcohol, such as vodka

**Method:** Toast the coriander seeds in a dry skillet over medium heat for 3–7 minutes, until they begin to smell aromatic. Coarsely grind the coriander seeds, celery seeds, lemongrass, nutmeg, cumin, and salt with a mortar and pestle. Transfer all the ingredients to an 8-oz (240-mL) jar, add the alcohol, and seal the jar with the lid. To infuse the ingredients, first shake the jar, then leave the jar to sit for 1 week, shaking and rotating it daily.

Strain the mixture through a nut milk bag or French press and discard the solids. Bottle the bitters.

Lemongrass

# Forager's Bitters

The bitters in this section will help you to start experimenting with a more diverse range of ingredients. We've indicated where you may need to forage a little, whether it is out in the wild, online, or in specialty stores that stock ingredients from all over the world.

## Lazy Hippie One-Day Bitters

This bitters is basically a scavenger hunt of the tea and tisane section of natural product grocery stores. These ingredients are prized for anticarcinogenic, antimicrobial, and anti-inflammatory capacity. You can use tea bags for most of the ingredients—each bag will contain about 2 g of dry herbs. Check the ingredients labels on the teas to choose brands with the fewest additives.

30 g hibiscus calyx

10 g pau d'arco bulk herb or tea

10 g chamomile tea

10 g jasmine green tea

10 g wild angelica bulk herb or tea

6 g licorice root bulk herb or tea

1¼ cups (295 mL) 75% ABV neutral grain alcohol

**Method:** Cut open any tea bags and empty the contents into a bowl. Put all the ingredients except the alcohol into a medium saucepan with 2 cups (475 mL) of water and bring to a boil, then simmer until the liquid is reduced by half and let cool. Strain the mixture through a nut milk bag or French press, then combine 1 cup (240 mL) of the mixture with the alcohol for a final ABV of about 40 percent. Bottle the bitters. Yields 500 mL.

## Danube Bitters

This blend celebrates the ecological and cultural diversity of the Danube, from the fruits of the region to the tradition of distilling fruit alcohol as *digestifs*.

1 handful fresh black currant berries and leaves

1 handful fresh blackberry berries and leaves

15 g dried black elderberry fruit

15 g dried unsweetened tart cherry fruit

15 g dried apricots

4 g paprika

2 g celery seed

15 g dried licorice root

10 g dried Hungarian hawthorn fruit

3 cracked walnuts with husk

7 g gentian root

2 cups (475 mL) 40% ABV grain alcohol, such as vodka

**Method:** Combine all ingredients in a 24-oz (700-mL) jar, add the alcohol, and seal the jar with the lid. Shake the jar to infuse the ingredients. Let the jar sit for 1 week, shaking and rotating it daily.

Strain the bitters through a nut milk bag or French press and discard the solids. Bottle the bitters. Yields 450 mL.

Jujube

## Alone in Kyoto Bitters

Jujube fruit, haw flakes, and shiso leaves can be bought in Japanese markets, and combinations of these ingredients often manifest in some of the highly meticulous and endlessly creative cuisines of the Kyoto Prefecture, particularly desserts and refreshments. Umeshu is a Japanese liqueur.

About 30 fresh pine needles

5 dried jujube fruit, separated into pieces

20 g fresh orange peel

30 g haw flakes

4 fresh shiso leaves

15 g peeled fresh ginger

2 cups (475 mL) 15% ABV Umeshu

1 cup (240 mL) 40% ABV grain alcohol, such as vodka

**Method:** Combine all the ingredients in a 24-oz (700-mL) jar, add the Umeshu and vodka, and seal the jar with the lid. To infuse the ingredients, first shake the jar, then leave the jar to sit for 1 week, shaking and rotating it daily.

Strain the mixture through a nut milk bag or French press and discard the solids. Bottle the bitters. Because this bitters has an ABV of about 23 percent, store it in the refrigerator. Yields 650 mL.

## Thai Market Bitters

All these ingredients can be found in any Thai grocery store. While most ingredients are from Southeast Asia, tamarind is from Africa and guava is from Central America. After 300 years, Thailand has its own distinct cultivars.

30 g dried coconut shavings

20 g fresh or dried galangal

5 kaffir lime leaves

45 g sweet tamarind fruit paste or 6 fruit, peeled and de-seeded

1 stalk lemongrass or 2 oz dried lemongrass tea

1½ cups (355 mL) 75% ABV neutral grain alcohol

60 g fresh guava fruit, chopped

15 g brown sugar

**Method:** Put the coconut, galangal, lime leaves, tamarind paste, and lemongrass into a 25-oz (750-mL) jar with the alcohol, and seal the jar with the lid. Shake the jar to infuse the ingredients. Let the jar sit for 1 week, shaking and rotating it daily.

Mesh strain the mixture to separate the liquids and solids and set the liquid aside. Transfer the solids to a medium saucepan with 2 cups (475 mL) of water and bring to a boil, then add the guava. Let simmer for 15 minutes, then add the sugar, remove from the heat, and let cool. Strain and combine 1 cup (240 mL) of the cooled water with the alcohol bitters, to bring the ethanol ABV to about 45 percent. Filter and let settle in a container overnight. Pour off the liquid into another container to avoid disturbing the settlement, which can be discarded. Bottle the bitters. Yields 500 mL.

## Biennial Bitters

Biennials are plants that do most of their flowering and fruiting every other year. We complemented the earthy flavor of the roots biennials with fruits and tarragon.

1 large carrot

2 raw medium beets

2 stalks celery

1 medium artichoke

1 medium fennel bulb

1 apple

1 pear

2 sprigs fresh tarragon or 2 g dried tarragon

1½ cups (355 mL) 75% ABV neutral grain alcohol

1 Tbsp granulated sugar

1 pinch sea salt

**Method:** Preheat the oven to 200°F (95°C). Juice the fresh ingredients and retain the solids (reserved by the juicer). Alternatively, you can blend, straining the blended juice through a mesh French press or nut milk bag. Enjoy the juice part fresh! Spread the blended solids and the tarragon on a baking sheet and bake for 2 hours with the oven door cracked open, if possible, to let the moisture out. Scrape 3½ oz (100 g) of the dehydrated solids into a 24-oz (700-mL) jar, add the alcohol, and seal the jar with the lid. Shake the jar to infuse the ingredients. Let the jar sit for 1 week, shaking twice daily.

Strain the bitters through a nut milk bag or French press and discard the solids. Boil 1 cups (250 mL) water, add the sugar and salt, and let cool. Combine ¾ cup plus 3 Tbsp (220 mL) of the boiled water solution with the alcohol bitters for a final ABV of about 45 percent. Bottle the bitters. Yields 500 mL.

## Chicha Morada Bitters

The purple maize that gives *chicha morada* its characteristic color is substituted here with carrots and hibiscus. This sweet, tangy blend will add color and zest to your drink.

1 sweet corn cob (fresh)

40 g purple carrots, peeled, sliced thinly

10 g dried roselle (hibiscus)

20 g maca powder

10 g dried pineapple fruit

5 clove buds

0.25 g cinnamon

20 g white granulated sugar

1½ cups (355 mL) 75% ABV neutral grain alcohol

**Method:** Grill the corn cob in a skillet or barbeque over medium heat, until kernels are light brown. Cut the corn kernels off the cob and discard the cob. Coarsely grind all the ingredients together, except the alcohol, using a mortar and pestle. Transfer to a 24-oz (700-mL) jar, add the alcohol, and seal the jar with the lid. To infuse the ingredients, first shake the jar, then leave the jar to sit for 1 week, shaking and rotating it daily.

Pour off the liquid and reserve. Transfer the solids to a medium saucepan with 2 cups (475 mL) of water, bring to a boil, and let simmer until reduced to two-thirds volume. Remove from the heat and let cool. Strain the water through a French press or nut milk bag, then combine ¾ cup plus 3 Tbsp (220 mL) of the cooled water with the alcohol bitters for a final ABV of about 45 percent. Bottle the bitters. Yields 500 mL.

## The Sitting Room: Oman Incense Bitters

In Oman, every part of the date plant is used, including the seed, which can be used to make a kind of coffee alternative. Most of the ingredients for this blend can be sourced from a Middle Eastern grocery store or online herb supplier. Ensure that the resins are 100 percent pure with no fillers.

7 dates

15 g frankincense resin

10 g myrrh resin

Zest from ½ fresh Key lime

9 g dried hyssop leaves

4 green cardamom pods

1½ cups (355 mL) 75% ABV neutral grain alcohol

2 tsp rose water

**Method:** Smash the dates and remove the seed from each, reserving both the fruit and the seeds. Wash the seeds and toast them in a dry skillet until they begin to smell aromatic. Transfer the dates and seeds to a 24-oz (700-mL) jar along with all the dry ingredients, then add the alcohol and seal the jar with the lid. Put the jar in a hot water bath (place the jar in a container of hot water, such as a pot in the sink) for 1 hour. Let the water cool, then remove the jar from the bath and let sit overnight.

The next day, boil ⅞ cup (200 mL) of water and let cool, then add the cooled water to the jar of bitters for a final ABV of about 45 percent. Add the rose water, then strain the mixture through a nut milk bag or French press. Bottle the bitters. Yields 500 mL.

Myrrh

## Mediterranean Garden Bitters

This bitters is a bright, fragrant blend from the fruits and spices of the Mediterranean region. Citrus, pungent, and savory, this simple blend can be made out of many plants typically found in a Persian or Turkish grocery store.

20 g coriander seeds

20 g dried pomegranate seeds or 15 g unsulfured raisins

5 g dried parsley leaves

1 dried bay leaf

30 g peeled dried pomegranate inner rind

10 g star anise

50 g dried lemon peel

Pinch of dried thyme

Pinch of dried rosemary

1½ cups (355 mL) 75% ABV neutral grain alcohol

**Method:** Peel away outer husk of pomegranate and set aside inner rind and seeds to sun-dry or use a dehydrator. Toast the coriander seeds in a dry skillet over medium heat for 3–7 minutes, until they begin to smell aromatic. Coarsely grind all the ingredients together using a mortar and pestle, transfer them to a 24-oz (700-mL) jar, add the alcohol, and seal the jar with the lid. Shake the jar to infuse the ingredients. Let the jar sit for 1 week, shaking and rotating it daily.

Pour off the liquid and reserve. Transfer the solids to a medium saucepan with 2 cups (475 mL) of water, bring to a boil, and let simmer for 10–15 minutes. Remove from the heat and let cool. Strain the water through a French press or nut milk bag, then combine ¾ cup plus 3 Tbsp (220 mL) of the cooled water with the alcohol bitters for a final ABV of about 45 percent. Bottle the bitters. Yields 500 mL.

## Tree Kisser (Eastern Forest Bitters)

Eastern Forest blend represented an attempt to bottle a familiar landscape during graduate school on the East Coast. A mixologist once told us it tasted like kissing a tree.

15 g paperbark birch bark or birch leaf tisane

10 g wild cherry bark or 20 ml wild cherry extract

10 g white oak bark or white oak bar tisane

7 g sassafras leaves or gumbo filé

15 g staghorn sumac seeds, or 5 g dried sumac berries or 2 g powdered sumac

8 g black walnut husks

6 g slippery elm inner bark

1 g black spruce resin

1 g Douglas fir resin or 20 g Douglas fir tips

15 g eastern red cedar leaves tips

1 Tbsp maple syrup

5 g wintergreen leaves or fresh spearmint leaves

8 g dried or 15 g fresh flowering dogwood fruits

8 g dried hawthorn fruit

1½ cups (355 mL) 75% ABV neutral grain alcohol

**Method:** Combine all the solid ingredients in a 24-oz (700-mL) jar, add the alcohol, and seal the jar with the lid. To infuse the ingredients, first shake the jar, then leave the jar to sit for 1 week, shaking and rotating it daily.

Pour off the liquid and reserve. Transfer the solids to a medium saucepan with 2 cups (475 mL) of water, bring to a boil, and let simmer for 10–15 minutes. Remove from the heat and let cool. Strain the mixture through a French press or nut milk bag, then combine ¾ cup plus 3 Tbsp (220 mL) of the cooled water with the alcohol bitters for a final ABV of about 45 percent. Pour the strained blend through coffee filters and collect the flow through. Bottle the bitters. Yields 500 mL.

Tamarind

## Veracruz Bay Bitters

Laurels are champion flavors in Central America cuisine, such as horchata and guacamole. This is a sweet bitters echoing candied cinnamon but with a summery twist.

2 sweet tamarind fruit, peeled and seeded, or 1 Tbsp tamarind paste

2 strips (20 g) apple fruit leather

Peel from 1 large avocado

3 bay leaves

2 g anise seed

5 g cassia cinnamon

Zest from 1 lemon

1½ cups (355 mL) 75% ABV neutral grain alcohol

1 vanilla bean or ½ tsp vanilla extract

**Method:** Combine all the solid ingredients in a 24-oz (700-mL) jar, add the alcohol, and seal the jar with the lid. To infuse the ingredients, first shake the jar, then leave the jar to sit for 1 week, shaking and rotating it daily.

Fine strain to separate the liquids and solids, separating both. Transfer the solids to a medium saucepan with 2 cups (475 mL) of water, bring to a boil, and let simmer for 10 minutes. Remove from the heat and let cool. Fine strain the mixture through a French press or nut milk bag, then combine about ¾ cup plus 3 tbsp (220 mL) of the cooled water with the alcohol bitters for a final ABV of about 45 percent. Strain the bitters through a nut milk bag or French press and discard the solids. Bottle the bitters. Yields 500 mL.

## Fruitbelt Bitters

We designed this bitters to pair with Fruitbelt Soda, a sparkling fruit drink designed to capture the prairies and heritage orchards of the American Midwest.

80 g dried heirloom apples, cored

25 g chicory root

13 g propolis resin

10 g aronia dried fruit

25 g dried ground dandelion root

1½ cups (355 mL) 75% ABV neutral grain alcohol

15 g malic acid powder

**Method:** Combine all the solid ingredients except the malic acid in a 24-oz (700-mL) jar, add the alcohol, and seal the jar with the lid. To infuse the ingredients, first shake the jar, then leave the jar to sit for 1 week, shaking and rotating it daily.

Pour off the liquid and reserve. Transfer the solids to a medium saucepan with 2 cups (475 mL) of water, bring to a boil, and let simmer for 10–15 minutes. Remove from the heat and let cool. Strain the water through a French press or nut milk bag, then combine about ¾ cup plus 3 tbsp (220 mL) of the cooled water with the alcohol bitters for a final ABV of about 45 percent. Add the malic acid powder and stir the bitters until the powder has completely dissolved. Strain the bitters through a nut milk bag or French press and discard the solids. Bottle the bitters.

Chicory

# Sipping Bitters

**These are recipes for grain alcohol or wine-base drinks that you can chill and drink straight in 1 to 2-oz (30–60-mL) volumes. They are less concentrated than cocktail bitters but can still be as potent.**

## Chewstick Bitters

Many of Manhattan's Dominican residents rely on botanicals as both a cultural resource and a form of healthcare. This bitters features several of the most common species found in the Dominican bitter infusion *mamajuana* and includes several species used traditionally for dental health, nicknamed "chewsticks." Serve with any rum-base drink.

15 g dried basil leaves

5 g dried Florida anise tree twigs and fruit

10 g dried cinnamon inner bark

3 g clove buds

15 g dried bay rum tree leaves

10 g dried timacle roots

10 g dried princess vine aerial parts

8 g dried minnieroot roots

8 g dried anamu roots

12 g dried licorice root

10 g dried China root

3⅛ cups (750 mL) 40–50% ABV rum (don't use overproof rum)

**Method:** Combine all dry ingredients in a 1-qt (1-L) jar, add the alcohol, and seal the jar with the lid. Shake the jar to infuse the ingredients. Strain the bitters through a nut milk bag or French press and discard the solids. Bottle the bitters. Yields 700 mL.

## Aloe Cooling Bitters Liqueur

You can add this bitters liqueur to any clear liquor, with citrus and herbs, such as more mint, for a cool-as-a-cucumber cocktail. It also goes well with melon juice.

1 large (at least 12 inches/30 cm) aloe leaf, spines removed

3⅛ cups (750 mL) 43% ABV pisco

1 medium horned melon, or 1 cucumber

15 fresh mint leaves, e.g. spearmint

Zest from 1 lemon

10 g ground orris root

½ cup (120 mL) rich syrup (see page 124)

**Method:** Cut the aloe lengthwise into several long pieces of gelatin, retaining as much as you want of the outside of the leaf. Put the aloe pieces and the pisco into a blender and blend for 1 minute, then transfer to a 1-qt (1-L) jar. Separate the inner gel and seeds from the horned melon rind and add to the jar, discarding the rind. Bruise the mint and add the leaves to the jar along with the lemon zest and orris powder. Seal the jar, then shake it to infuse the ingredients. Let the jar sit for 1 week, shaking and rotating it daily.

Strain the bitters through a metal mesh strainer and then a nut milk bag or French press. Add the rich syrup to finish. Yields 800 mL.

Mint

## Amphora Dry Vermouth

Early version of vermouths often had wormwood, while others had tree resins. Here, we include both.

10 g dried gentian root

5 g dandelion dried root

15 g lemon verbena dried leaves

30 g wormwood dried herb

10 g juniper cones

10 g dried Egyptian chamomile

5 whole coriander fruit

15 black peppercorns

3 green cardamom pods

5 g ground cinnamon

2¾ cups (650 mL) 14.5% ABV Trebbiano white wine

10 g pine, fir, mastic, or other tree sap crystals, depending on your preference

Zest from 1 lemon

1⅝ cups (375 mL) 20% ABV Palo Cortado sherry or other dry sherry to increase the wine to about 16.5% ABV

**Method:** Put all the ingredients except the sap crystals, lemon zest, and sherry into a medium saucepan, bring to a boil, and lightly simmer for 8 minutes. Remove from the heat and let cool. Transfer the mixture to a 1-qt (1-L) jar, add the sap crystals and lemon zest, seal with the lid, and let it sit overnight.

Strain the bitters through a nut milk bag or French press and discard the solids, then add the sherry. Bottle the bitters; keep them refrigerated and use within a month. Yields 800–900 mL.

## Hazel's Formula Sweet Vermouth

This vermouth is herbaceous and complex, but one without a dominant flavor. Don't sweat if you decide to swap an ingredient. Hazel is Rachel's rambunctious pit bull, who is just as forgiving as the recipe. On a geeky note, all the botanical ingredients rely on animal pollination, mainly either by bees or by humans.

10 g dried gentian root

20 g dried angelica root

1 vanilla bean

20 g fresh orange peel

10 g mahlab cherry seeds

8 g clove buds

2 mashed dates

10 g dried honeysuckle

10 g dried yarrow flowers and leaves

15 g dried bee balm

10 g mugwort

5 sage leaves

1 bay leaf

10 coriander fruit

1 small pinch saffron

5 rose buds

3⅛ cups (750 mL) 14.5% ABV Trebbiano white wine

200 g (1 cup) sugar

1 peach, including the pit, sliced (note that this won't be needed until 5 days after the first stage of the recipe)

⅞ cups (200 mL) 35% ABV brandy or eau-de-vie, to increase the wine to about 16.5% ABV

**Method:** Coarsely grind together all the ingredients except the peach, sugar, wine, and brandy using a mortar and pestle. Transfer to a 1-qt (1-L) jar, add the wine,

then seal the jar with the lid. Shake the jar to infuse the ingredients. Let the jar sit for 1 week, shaking and rotating it daily.

After 5 days, put the sliced peach and pit in a separate jar. Warm the brandy on low in a microwave for 1 minute, then it pour over the peach and seal with the lid. Let this mixture age together for 2 days (continuing to shake and rotate the wine mixture each day), then strain it through a French press or nut milk bag while cutting and mashing the peach with a knife to extract the juice. Reserve the brandy mixture and throw away the peach solids.

To caramelize the sugar, pour it in an even layer in a heavy skillet or saucepan, then add ¼ cup (60 mL) water, moistening all the sugar. Warm it over low heat, stirring constantly with a wooden spoon. When it looks like a sandy slurry, stop stirring and increase the heat, and heat until the mixture comes to a low boil. When the color is golden, remove the skillet from the heat, and slowly stir in ⅞ cup (200 mL) of the brandy mixture. Add the caramel–brandy mixture to a new 1-qt (1-L) jar. Strain the bitters through a nut milk bag or French press and discard the solids. Finally, combine the brandy with the wine, tasting as you do so to obtain the desired sweetness. Store in the refrigerator. Yields 1 L.

Gentian

# Bitters Masterclass

This is a how-to guide on how to make a traditional bitters in eight steps. It is the method we use for all our signature bitters (see pages 64–68) and our recipes for re-creating classic bitters (see pages 68–69). You can also use it for creating your own bitters recipes (see pages 162–65).

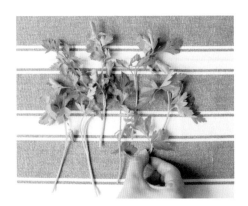

### Step 1: Cleaning and Drying
If using fresh plant material, first clean it by rinsing in water and then dry it at room temperature in a well-aerated space in the shade, either indoors or outdoors. If using dried plant material that you purchased, spread it out on your work space and check for any impurities or contamination.

### Step 2: Grinding
Reduce the botanicals into a fine powder using a mortar and pestle or a spice grinder. Some botanicals are difficult to grind up, in which case larger pieces will have to suffice.

If making your own recipe, about 50 g (1¾ oz) of dried plant material should be enough for simple combinations of ingredients; more complex recipes may require 100 g (3½ oz) or more. Weigh your botanical material using a kitchen scale.

### Step 3: Mix with the Solvent

Put the botanicals into a glass jar, then pour in the alcohol solvent. Use a strong alcohol, such as 75% ABV. The jar should be filled to the neck to minimize the surface being in contact with oxygen, but at the same time the botanical should avoid direct contact with the lid.

If making your own recipe, pour over 1⅝ cup (375 mL) of the selected solvent for every 1¾ oz (50 g) of ground plant material.

### Step 4: Get Shaking and Infusing

Perturb the botanical-and-solvent mixture by shaking the jar. You will probably notice a change in the color of the solvent after you have infused your dried botanical.

If you're not intending to use your bitters immediately, you can enhance the infusion by storing the jars upright in a cool, dark place (light can degrade the pytochemicals) for a week. Shake and rotate the jars each day. To speed up this infusion step, see sonication method on page 165.

### Step 5: Filter Your Infusion

Pour the extracted botanical infusion mixture into a French press and press to filter the extract, or alternatively squeeze the infusion through a nut milk bag. You may need to repeat this step several times for the desired clarity. The resulting liquid is the concentrated alcohol extract; set it aside while you use the solids in step 6.

## Step 6: Mix In the Water

To create the water extract, add the solids to a large stockpot and cover with about 1⅔ cups (400 mL) of water. Simmer over low heat for around 15 minutes—be careful because splashed ethanol can catch fire. Remove from heat and let cool. Strain through a fine mesh strainer or a nut milk mesh bag, then discard the solids.

If you are making your own recipe, pour over about four times as much water as you have solid material.

## Step 7: Dilute

Mix 1 cup (240 mL) of the water extract with 1¼ cups (295 mL) of the alcohol extract (calculated based on 75% ABV alcohol). The diluted extract can become cloudy, because compounds in the alcohol that are not soluble in the water extract fall out of solution. You can filter it again using a paper coffee filter.If you are making your own recipe, use the formula on page 163 to calculate the amount of water and alcohol extracts to combine for your desired ABV.

## Step 8: Bottle and Store

Pour the final extract into clean glass bottles. We prefer to use lids with glass droppers. Be sure to include the date and storage directions on the label. We also encourage writing all the ingredients on the label.

Bitters are best stored in a cool, dry place, out of direct light and heat, both of which oxidize and degrade phytochemicals. Alcoholic bitters with a high alcohol content (>40% ABV) if kept properly sealed can generally be stored for one year or longer.

# Shoots & Roots House Recipes

The recipes in this section are Shoots & Roots house recipes, all created using our standardized bitters-making method, using 75% ABV neutral grain alcohol and ending with 42% ABV bitters.

## Summers in Bangkok Bitters

Imagine childhood summers gorging on tropical fruit, running around food markets and temples, blissful days of breathing in rich experiences that seem to never end. This is the experience Selena had that she infuses in this recipe. Many of the sweet-and-sour flavors of Thailand also have antimicrobial properties, so there's more than one reason to savor them.

25 g dried bael fruit

7.5 g dried blue pea flowers

27.5 g mangosteen powder

15 g galangal

40 g dried amla fruit

3 fresh kaffir lime leaves

½ kaffir lime fruit, fresh or dried

22.5 g dried hibiscus flowers

2 Tbsp sour tamarind fruit paste

1 Tbsp turmeric powder

1 fresh or frozen pandan leaf

35 g sliced reishi mushrooms

1⅝ cups (375 mL) 75% ABV neutral grain alcohol

Method: Use the standard bitters-making process for making house recipes on pages 60–63. Yields 500 mL.

## Demon Flower Bitters

These twelve plants hail from Mexico and the Philippines and symbolize the deep cultural trade of medicinal and ritual plants rooted in the colonization of New Spain.

25 g Devil's dried hand flowers

25 g pau d'arco bark

12.5 g hibiscus flowers

5 g dried sambong leaves

10 g dried uchuva fruit

2.5 g dried jasmine flowers

5 g dried chamomile flowers

5 g dried blue skullcap leaves

0.5 g dong quai root

15 g cacao seeds

2 g dried blue pea flowers

7.5 g dried linden flowers and leaves

1 g propolis extract

1⅝ cups (375 mL) 75% ABV neutral grain alcohol

Method: Use the standard bitters-making process for making house recipes on pages 60–63. Yields 500 mL.

## Black Bear's Bitters

Several of the First Nations credit bears for leading them to the powerful medicinal osha root. Bears use these plants for sustenance, pain relief, and to aid digestion. Magnolia bark is sold under its Chinese name, Hou Po.

27.5 g dried labrador herb

5 g thin osha (*Ligusticum grayii*)

25 g dried fat osha root (*Ligusticum porterii*)

15 g dried elderberry fruit

15 g juniper berries

7.5 g sloe fruit

3.5 g dried sasparilla

7.5 g dried pansy leaves

10 g dried dandelion leaves

7.5 g dried passionflower leaves and stems

15 g magnolia bark tea

5 g prickly ash bark

5 g prickly ash seeds

8.5 g cranberry fruit, preground

1⅝ cups (375 mL) 75% ABV neutral grain alcohol

**Method:** Use the standard bitters-making process for making house recipes on pages 60–63. Yields 500 mL.

Labrador

Amaranth

## Andean Amargo Bitters

We present ancient roots and grains cultivated by the Aymara, Quechua, and Incans in the high-elevation Andean ranges of Peru. It's not a bitters for the fainthearted; maca has purported aphrodisiac properties, and hercampuri are the bitterest roots of the New World.

50 g powdered maca root

27.5 g dried lemon verbena leaves and stems

2 Tbsp chia seeds

1 Tbsp quinoa seeds

1 Tbsp amaranth seeds

20 g entire dried lungwort plant

10 g entire canchalagua plant, dried

2.75 g entire hercampuri plant, dried

57.5 g lucuma fruit powder

1⅝ cups (375 mL) 75% ABV neutral grain alcohol

**Method:** Use the standard bitters-making process for making house recipes on pages 60–63. Yields 500 mL.

Tamarind

## Mexican Molcajete Bitters

*Molcajete* is the name for a traditional Mexican mortar and pestle, derived from the Náhuatl *molcaxitl*. This blend invokes the aromas and spices of plants and sacred earth used in traditional Mexican food and medicine.

225 g sweet tamarind fruit

7.5 g dried gordolobo inflorescences

3.5 g dried avocado leaves

3.5 g dried hoja santa leaves

2 g dried epazote leaves

2 g pericon (*marigold*) shoots

1 large cinnamon stick

½ vanilla bean or ½ tsp vanilla extract

37.5 g blend of the following chili pepper fruits: Chipotle Meco, Chipotle Morita, Chili Pulla, Chili Costeño, Chili De Arbol, Chiltepin

1 pinch tierra santa (kaolinite clay), add after dilution

1⅝ cups (375 mL) 75% ABV neutral grain alcohol

**Method:** Use the standard bitters-making process for making house recipes on pages 60–63. Yields 500 mL.

## Hanging Gardens of Babylon Bitters

This is a blend inspired by the elusive Wonder of the Ancient World, and a reimagining of the Mediterranean and Middle Eastern herbs that may have grown there. Was this legendary garden a mythical ideal or a reality, and what may have been cultivated?

20 g ground pomegranate seed and fruit

30 g lightly toasted black barley; do not pulverize (keep whole)

15 g coriander seeds, toasted

20 g dried parsley root

10 g dried parsley leaves

7 bay leaves

5 g star anise

10 g sumac fruit, fresh or dried

10 g dried amla fruit

5 g dried bergamot leaves

10 g dried lemon peel

2.5 g thyme herb

1⅝ cups (375 mL) 75% ABV neutral grain alcohol

**Method:** Use the standard bitters-making process for making house recipes on pages 60–63. Yields 500 mL.

Pomegranate

Mume plum

## Ki Japanese Arboreal Bitters

This bitters design started as a project for Mizu Shochu, because most market bitters overpowered shochu notes. Ki is sultry, earthen, fruity, and floral, invoking the botanist in all of us to study the magnificent aromatic arbors. The plums shouldn't be blended, because the pits can break your grinder or blender. Instead, we suggest puncturing them with a knife to get them to release their flavor faster.

100 g smoked dried Japanese mume (ume) plum fruit

20 g dried Japanese allspice flower buds

15 g eastern white cedar branches

12.5 g dried hawthorn fruit

10 g dried jujube fruit

10 g kousa dogwood fruit, leaves, and twigs

¼ luo han guo fruit for sweetness, added only while making the water extract

2.5 g propolis resin

1⅝ cups (375 mL) 75% ABV neutral grain alcohol

**Method:** Use the standard bitters-making process for making house recipes on pages 60–63. Yields 500 mL.

## Chai Jolokia Bitters

Naga jolokia is a north Indian variety of chili pepper once considered the hottest in the world. The native Indian crops in masala chai are used in Ayurveda to stimulate appetite and boost immunity.

30 g black tea leaves

5 green cardamom pods

3 dried black cardamom pods

10 g ground cinnamon

15 g dried ginger rhizome

10 g dried mace aril

5 g nutmeg seed

5 g (7 sticks) long pepper

5 g black pepper

4 g clove buds

10 g star anise

1 tsp ground naga jolokia pepper

1⅝ cups (375 mL) 75% ABV neutral grain alcohol

**Method:** Use the standard bitters-making process for making house recipes on pages 60–63. Yields 500 mL.

Bael fruit

### Mount Apo Bitters

Our Mount Apo bitters is made from an herbal tea blend, Pito-Pito, which is common in the Philippines. This blend of seven different plant types (banaba leaf, guava leaf, pandan leaf, lemongrass stalk, anise, sambong leaf, and moringa leaf) is high in antioxidants. Rich in vitamins, this is a green, savory bitters that revitalizes the spirit.

4 g ground black walnut

2.5 g dried rue leaves

2.5 g yohimbe bark

5 g dried sambong leaves

5 g dried orange peel

5 g dried lemon peel

5.5 g star anise

6 g dried moringa leaves

6 g dried mugwort leaves

10 g dried lemongrass stalk

5 g dried banaba leaves

7.5 g dried guava leaves

7.5 g dried mango leaves

½ large fresh or frozen pandan leaf

1 g dried orris root

Light pinch salt

1⅝ cups (375 mL) 75% ABV neutral grain alcohol

**Method:** Use the standard bitters-making process for making house recipes on pages 60–63. Yields 500 mL.

Banaba

# Classic Bitters

**Most cocktail recipes call for three kinds of bitters: Angostura, Peychauds, and orange bitters. Here are our recipes. We also sell the herbs as ready-to-extract blends.**

### Raisin in the Sun Bitters

This is the homemade version of Peychaud's bitters, which was meant to be enjoyed in a cognac Sazerac but now has a whiskey base because of the Phylloxera insect that invaded and wiped out most of the grape plantations. So we added a little grape to this recipe in place of sugar. Don't try to blend the raisins; instead, cut each in half with scissors and add them to the alcohol to infuse.

60 g raisins, cut in half

20 g dried pomelo peel

10 g bitter orange peel

10 g lime peel

20 g celery seeds

10 g fennel seeds

1 whole star anise

10 g gentian root

8 g allspice fruit

20 g dried sloe fruit

7 g dried juniper berries

5 g ground nutmeg

5 g ground cinnamon

1⅝ cups (375 mL) 75% ABV neutral grain alcohol

**Method:** Use the standard bitters-making process for making house recipes on pages 60–63. Yields 500 mL.

## Citrus Grove Bitters

We've been asked by many bartenders to make a better citrus bitters, and this is it. We think a mixture of plant parts and combination of fresh and dried ingredients creates the full-spectrum bitters we've been searching for.

30 g fresh or 20 g ground dried turmeric rhizome

1 large fresh ginger rhizome

10 g dried bitter orange fruit peel

40 g dried orange fruit peel

40 g fresh grapefruit peel

40 g fresh tangerine peel

30 g dried whole or sliced kumquat fruits

15 g dried licorice root

5 green cardamom pods

1 fresh kaffir lime leaves (optional for floral fan)

2 to 5 drops orange blossoms or ¼ tsp orange blossom water

2 cups (480 mL) 75% ABV neutral grain alcohol

¾ cups (150 g) sugar

**Method:** Use the standard bitters-making process for making house recipes on pages 60–63, using all the ingredients except the sugar. Add 240 mL of the water extract to the full alcohol extract volume. After mixing the water and alcohol extracts together at the end of step 7, dissolve the sugar in the mixture. This ABV is higher. Yields 650 mL.

## Angostura Trifoliata Bitters

Some of these ingredients for this recipe are hard to procure. We suggest you try working with what you have available.

60 g canella bark

30 g dried bitter orange zest

30 g angostura bark

25 g ground tamarind

20 g gentian root

20 g dried lemon zest

20 g allspice fruit

20 g clove buds

20 g dried ginger rhizome

20 g mauby bark

20 g dried simarouba leaves or bark

20 g sweetbroom

20 g puron bark

10 g balsam leaves

10 g mace aril

15 g dried licorice root

5 g star anise

2 green cardamom pods

2 charred white oak sticks (left whole)

2 shaved tonka beans

2 cups (480 mL) 75% ABV neutral grain alcohol

½ cups (100 g) sugar

**Method:** Use the standard bitters-making process for making house recipes on pages 60–63, using all the ingredients except the sugar. Add 240 mL of the water extract to the full alcohol extract volume After mixing the water and alcohol extracts together at the end of step 7, dissolve the sugar in the mixture after bringing the bitters to 46, not 42, percent ABV. Refrigerate the bitters and shake immediately before use. Yields 650 mL.

Chapter 5
# Bitters at the Bar

You can make your own delicious cocktails with flavors from around the world. In this chapter, we divulge our most coveted recipes for cocktails and syrups, created by us and renowned mixologists Christian Schaal and Kevin Denton. Each cocktail recipe is paired with a botanical profile featuring a plant in the recipe. There are also ideas for using bitters in other recipes, from coffee to ice cream. Next, dive into a revived element of the bar: the shrub. Recipes created by Jim Merson will jolt the physiology of fruit into long-term memory. Our hope is that making these recipes heightens your awareness and enjoyment of plants.

# Tools of the Trade

The cocktails presented here can be made with kitchen equipment that you probably already have on hand. However, you may want to invest in a few basic bartender's gadgets to make following the cocktail recipes an easier and more fun experience. Our own arsenal of tools for making cocktails is usually simple, although sometimes we enjoy technology by making drinks using liquid nitrogen, centrifuges, cream whippers, dry ice, and sonicators. But don't worry—you don't need any of these to make great drinks.

The typical kitchen equipment we use for making cocktails includes a sharp knife and cutting board for cutting and slicing various ingredients, a vegetable peeler for creating garnishes, and an ice cube tray for making ice cubes. You can manually juice citrus by hand, but a juicer or citrus press will take the work out of juicing. A tradtional juicer with a ridged, dome shape centered on a saucer works well for smaller amounts.

We use a jigger for measuring alcohol. This bartender's measuring cup often has an hour-glass shape and holds 1 ounce on one end and 1½ ounces on the other, but these measurements can vary.

A cocktail shaker is ideal for mixing drinks with juice or egg and some have a built-in strainer. Stir drinks that are made of alcohol and syrups. Shaking a drink is a great way to mix and open up the ingredients while chilling them, if shaken with ice. If you have

one without a strainer, look for a hawthorn or julep strainer, which are designed for straining drinks from a shaker. A bar spoon with a long handle and a mixing glass (any spouted vessel) are useful for stirring drinks. A muddler has a rounded end useful for crushing botanicals in the bottom of a glass, but you can use the end of a wooden spoon instead.

## A NOTE ON MEASUREMENTS

Because ounce-measuring jiggers are often used to conjure up cocktails, the recipes in this chapter provide measurements based on ounces. If you don't have a jigger, you can convert the measurements to tablespoons; 1 ounce equals 2 tablespoons. A bar spoon holds ½ teaspoon.

# Glasses

We like to switch up glasses depending on the cocktail being made. The basic cocktail glasses we have on hand are highball glasses, old-fashioned glasses, and coupes. Glassware choices are based on whether ice should be in the drink, the volume of liquid, how much you want to direct effervescence or aroma, and plain old aesthetic. We also like to play around with serving drinks in other vessels, such as ceramic cups or inside fruits. For example, when we ran an event on agave for the Museum of Food and Drink, we served cocktails in banana peppers.

### Highball Glass
One of the more common cocktail glasses, a highball glass is tall with straight sides, perfect for holding long drinks made with mixers and ice.
8–10 oz (240–300 mL)

### Collins Glass
A Collins glass is a little narrower and taller than a highball—these can be used interchangeably. A taller version that holds 16 ounces (475 mL) can also be found.
10–14 oz (300–410 mL)

### Old-Fashioned Glass
Also known as a rocks glass or lowball glass, this glass takes its name from the classic old-fashioned cocktail. It is typically used for drinks that are served over ice but with little or no mixers.
6–10 oz (180–300 mL)

### Double Old-Fashioned Glass
Often abbreviated to DOF, this is simply an oversize old-fashioned glass. Its size and broad rim allow for ingredients to be crushed directly inside it.
12–16 oz (350–470 mL)

### Cocktail Glass/Martini Glass
Visually similar, the cocktail glass has slightly rounded sides while the martini glass, which developed from it, is conical. The martini glass has a longer stem to hold, helping to keep the drink cool.
3–6 oz (90–180 mL)

### Coupe Glass
Though originally used for drinking champagne, the coupe glass is now more often used for any cocktail served without ice. The coupe glass has a rounded saucer-shaped bowl, which reduces the chances of spilling its contents or knocking it over.
5 oz (150 mL)

| Champagne | Martini | Coupe | High ball | Old-fashioned | Double old-fashioned | Shot | Snifter |

### Margarita Glass
Based on the coupe glass, the wide bowl of this glass makes it easier to add salt to the rim for a margarita cocktail.
6–8 oz (180 to 240 ml)

### Shot Glass
A small sturdy glass that is too small for most drinks, we use it for the Dia Siete cocktail (see page 105). A shot glass that holds 1 ounce (30 mL) can double as a jigger.
1–2 oz (30–60 mL)

### Champagne Flute
The narrow bowl of this tall glass help to prevent bubbles escaping.
6 oz (180 mL)

### Wine Glass
This glass has a large bowl suitable for swirling wines. Stemmed versions are best for white wines while stemless ones are ideal for reds.
6–14 oz (180–410 mL)

### Beer Glass
These vary in shape and size, from beer mugs with a handle to pint glasses and goblets to V-shaped pilsner glasses.
10–20 oz (300–600 mL)

### Snifter
A short-stemmed glass with a globular bowl, the snifter is perfect for swirling brandies.
6–8 oz (180–240 mL)

---

## PERFECT ICE

We get pretty excited when using large, clear ice cubes for making cocktails. The size of the ice cubes is important. Small ice cubes melt faster compared to large ones, because the cold ice of the small cubes have greater interaction with the warm drink and can dilute your cocktail faster than large ice cubes. So, when choosing ice cube trays, look for ones that will make large ice cubes. Clear ice cubes are beautiful and can trick the eye with suspended leaves and flowers in them. We make clear ice cubes by using hot water poured into an ice cube tray instead of cold or room temperature water. Just as important, the water you use to make ice cubes can affect the flavor of your drink. Some tap water is treated with chemicals that affects its flavor. If you want the ice cubes you make to have a neutral taste, use bottled, boiled, or filtered spring water.

# Tea

**Latin Name:** *Camellia sinensis* (L.) KUNTZE

**Botanical family:**
Theaceae (the Tea family)

**Common name(s):**
Tea, cha, chai

**Habitat:**
The native tea-growing area, or "tea belt," encompasses southwestern China, northern Laos, northern Vietnam, Myanmar, Cambodia, and northeastern India. The tea plant naturally grows in the understory of broad-leaved evergreen forests, below the canopy but above the forest floor.

**Parts used:**
Leaf, bud, flowers

**Diversity:**
There are more than 1,500 named cultivars and countless locally adapted varietals, also called landraces, of the tea species.

**Botany:** In its native form, tea grows as an elegant tree up to 60 feet (18 m) high; in most cultivated landscapes it grows as a shrub. It has distinct yellow-white flowers 1 to 1⅜ inches (2.5–3.5 cm) in diameter with six to eight petals. The leaves are serrated with a distinct webbing pattern. Young leaves have short white hairs on the underside.

**Cultural Uses:** Buddhists and Zen monks drank tea to keep themselves awake during long meditation sessions. A nice cup of tea is also, of course, a popular drink in many cultures. Today, tea is the most widely consumed beverage in the world besides water.

**Culinary Uses:** Commercial tea is primarily produced from *Camellia sinensis* var. *assamica*, or the broad-leaved variety of the plant, and *Camellia sinensis* var. *sinensis*, or the small-leaved variety of the tea plant.

**Traditional Medicinal Uses:** Tea is used to strengthen the immune system; balance the body's hot and cold levels; detoxify blood; treat rheumatism, stones, and headaches; and reduce swelling in soft tissue. It is used to treat mental well-being, invigorate the mind, and relieve stress as well as to provide nutrition, aid digestion, and prevent obesity.

**Phytochemistry and Bioactivity:** The claimed health benefits of green tea are mostly due to catechin polyphenols. Caffeine contributes to green tea's stimulant properties, while amino acid theanine contributes to its relaxing properties. Tea has more than five hundred volatile compounds that contribute its aroma.

**Safety Precautions:** Should be consumed with due care by people with caffeine sensitivity.

# Green Glory

By the Shoots & Roots team

Tea, sugar, and alcohol are easy to pair. We love the froth and flavor of Japanese matcha combined with oolong. The matcha becomes creamy and vegetal, and it imbues the cocktail with a bright, glowing green. High-grade oolong tea will leave a sweet aftertaste in your throat and make you salivate a little. This cocktail is full of both caffeine and L-theanine from the tea, as well as the B vitamins from the moringa plant that's in our Mount Apo Bitters.

---

⅛ tsp matcha powder

2 oz (60 mL) gin

¾ oz (20 mL) Oolong Tea Simple Syrup (page 125)

2 oz (60 mL) tonic water

5 drops Mount Apo Bitters (page 68) or Tree Kisser Bitters (page 56)

**Method:** Add the matcha powder, gin, and oolong to a shaker, shake to froth, strain, pour over ice in a snifter, and top with the tonic and drops of bitters. Garnish with the edible leaves of your choice. Yields one 4 oz drink.

# Ginger

**Latin Name:** *Zingiber officinale* ROSCOE

**Botanical family:**
Zingiberaceae
(the Ginger family)

**Common name(s):**
Ginger, common ginger

**Habitat:**
Widely cultivated in the tropics and subtropics, ginger is native to tropical Southeast Asia and possibly India, where the largest genetic diversity for the species exists.

**Parts used:**
Rhizome

**Diversity:**
There are 218 named species of *Zingiber*.

**Botany:** The shoots of ginger are pseudostems that arise annually from buds on the rhizome. Ginger flowers are cone-shaped spikes with greenish to yellowish leaflike bracts and pale yellow flowers that protrude just beyond the edge of the bracts.

**Cultural Uses:** It was one of the first Asian spices brought to Europe. It is widely used in a variety of traditional drinks.

**Culinary Uses:** Raw ginger is chopped or grated and used in soups, stir-fries, and meat, poultry, and seafood dishes across Asia; crystallized ginger root is a popular candy. The powder is popular in Western cuisine. The essential oil is used to flavor beverages.

**Traditional Medicinal Uses:** Ginger is a key ingredient in herbal medicines for treating rheumatism, nervous diseases, gingivitis, stroke, constipation, and diabetes. It is used against nausea and vomiting associated with vertigo, motion sickness, and morning sickness. Topically, the essential oil can be applied as an analgesic.

**Phytochemistry and Bioactivity:** Gingerol and its derivatives are reported to be more effective than aspirin in preventing blood clotting from platelet aggregation. Ginger has been proven to inhibit genes involved in the inflammatory response.

**Safety Precautions:** Ginger is generally considered a safe herbal medicine, but it may interact with the anticoagulant drug warfarin and the cardiovascular drug nifedipine. Its safety for women during pregnancy has not been established. It may cause heartburn and in high doses can be a gastric irritant; inhalation of dust from ground ginger can cause allergies. Species of *Asarum* have similar aromatic properties but contain aristolochic acid, a carcinogenic compound.

# Bodhi Tree

By the Shoots & Roots team

As the first Buddha Siddhartha wandered through India looking for a place to meditate, it was said he considered a grove of mango trees before settling under the shade provided by the large leaves of the Bodhi tree. If Siddhartha went wandering through Florida today, he would find a suitable spot at the USDA Subtropical Horticultural Research Station, which protects a genetic treasure-house of global importance: nearly three hundred accessions of mango, and twenty-three species of figs, including towering Bodhi trees. Whether this cocktail causes you to contemplate enlightenment or the value of safeguarding biodiversity for food and agriculture, we are sure you will find it refreshing.

---

2 oz (60 mL) Ginger-Infused Rum (see right)

3 dashes Chai Jolokia Bitters (see page 67) or Raisin in the Sun Bitters (see page 68)

¾ oz (20 mL) Mango Black Tea Syrup (see page 124)

½ oz (15 mL) lime juice

**Ginger-Infused Rum:** Infuse 8 thin slices of ginger in 6 oz (175 mL) rum in a glass jar at 86°F (30°C) for 30 minutes. Age 6 oz (175 mL) rum with 10 thin slices of ginger for 2 weeks in an airtight container, shaking briefly every other day to make sure the ginger is getting exposed to the rum. Store the container at room temperature away from light.

**Method:** Combine all the ingredients in a shaker and add ice. Shake and double strain, then garnish with a bruised kaffir lime leaf or a slice of mango. Yields one 4 oz drink.

# Mume Plum

**Latin Name:** *Prunus mume* (Siebold) Siebold & Zucc.

**Botanical family:**
Rosaceae (the Rose family)

**Common name(s):**
Ume, mei flower, Chinese plum, Japanese apricot, umboshi plum

**Habitat:**
Native to montane forests in China (western Sichuan and western Yunnan), Korea, Japan, Taiwan, northern Laos, and northern Vietnam at altitudes between 5,575 and 10,175 feet (1,700–3,100 m).

**Parts used:**
Fruit and flowers, whole trees for ornamental use

**Diversity:**
There are more than three hundred cultivars of the mume plum.

**Botany:** Mume trees generally grow up to 33 feet (10 m) and can grow as a shrub in certain environments. The aromatic flowers bloom in winter and spring, when the trees are leafless and can tolerate freezing temperatures. The fruit is ¾ to 1¼ inches (2–3 cm) in diameter, and it ripens during the rainy season.

**Cultural Uses:** The beautiful blossoms of the mume plum tree have been central to the art and poetry of China, Korea, and Japan since the Tang dynasty. Plum is one of the Four Noble Plants in Chinese culture, along with bamboo, orchid, and chrysanthemum. These plants are popular features as a bonsai and in special gardens.

**Culinary Uses:** Mume plums are widely used in liquor, juices, and sauces as a flavoring and nutritional enhancement and are smoked, pickled, and eaten as a snack. The plant's blossoms are used as a flavoring.

**Traditional Medicinal Uses:** Long used in traditional Chinese medicine to settle the stomach, resolve phlegm, and soothe the liver, mume plumes are prescribed to treat indigestion, stomachache, poor appetite, chest pain, and dizziness. Fermented mume plums contain probiotics, which boost the immune system.

**Phytochemistry and Bioactivity:** Studies in an animal model found unripe mume plums to have antimicrobial properties and to inhibit *Helicobacter pylori*, which is associated with gastritis and gastric ulcers.

**Safety Precautions:** Similar to other plums, almonds, and peaches, the seed of mume plums can produce hydrogen cyanide (the compound that gives almonds, *Prunus dulcis*, their distinct flavor). Hydrogen cyanide is generally present in too small a quantity to be considered toxic. As a safety precaution, the seeds should not be casually eaten.

# Wu Wei Mei

By the Shoots & Roots team
and Christian Schaal

It's smoky, it's sour, it's bitter, it's sweet, it's salty, it's warm, and it's earthy. Chinese five-spice powder (*Wu Wei*) is actually made of up to eight spices. Mume plums (*Wu Mei*) are a popular traditional medicine. They are genetically closer to apricots than plums.

---

1 oz (30 mL) Mume Plum Juice (see right)

1½ oz (45 mL) bourbon

½ oz (15 mL) mezcal

½ oz (15 mL) Chinese Five-Spice Syrup (see page 125)

¼ oz (7.5 mL) lemon juice

1 dash Ki Japanese Arboreal Bitters (see page 67) or Alone in Kyoto Bitters (see page 53)

Lemon peel and Himalyan blackberries (for garnish)

**Mume Plum Juice:** Simmer 12 oz (355 g) smoked plums in a large stockpot with 1 cup (240 mL) unsweetened prune juice, 1 cup (240 mL) water, and 1 cup (200 g) sugar for about 20 minutes, until soften a little. Let sit for 2 hours. Strain through a fine metal strainer and refrigerate in an airtight container for up to 10 days. Yields about 2 cups (475 mL).

**Method:** In a shaker, combine all the ingredients except the Ki Bitters and shake with ice. Double strain into a rocks glass over large ice cubes, then add the Ki Bitters. Garnish with lemon peel and some Himalayan blackberries. Yields one 4 oz drink.

# Schisandra

**Latin Name:** *Schisandra chinensis* (TURCZ.) BAILL.

| **Botanical family:** | **Common name(s):** |
|---|---|
| Schisandraceae (the Schisandra family) | Five-flavor fruit, magnolia-vine, Chinese magnolia-vine, omija, omija cha, gomishi, repnihat, wuweizi, lemonwood, Chinese magnolia vine, matsbouza |

**Habitat:**
A deciduous, woody vine (liana) native to northeastern China, Japan, and Korea, occurring naturally in mixed conifer–broad-leaved forests, valleys, and stream banks, growing on the trunks of small trees and bushes.

| **Parts used:** | **Diversity:** |
|---|---|
| Fruit (berry) | There are about sixteen species of *Schisandra* with berries used for medicine. |

**Botany:** Schisandra is the berry of a climbing vine. Flowers on a female plant will produce fruit when fertilized with pollen from a male plant. Male flowers have five stamens and anthers clustered on the receptacle; this structure has an unusual shieldlike form. Mature plants require sunlight at blossoming and fruiting stages.

**Cultural Uses:** Schisandra berries have a remarkable flavor profile reflected in its Chinese name, translated as "five-flavor fruit." Indigenous Nanai hunters used Schisandra for a variety of functions, including to improve night vision, as a tonic, and to reduce thirst and exhaustion.

**Culinary Uses:** Schisandra fruit are harvested for the commercial production of extracts, juices, wines, confectionaries, and teas, primarily in China and Russia.

**Traditional Medicinal Uses:** According to Chinese folklore, schisandra can "calm the heart and quiet the spirit." It is considered one of the fifty fundamental herbs of traditional Chinese medicine. Schisandra has been traditionally used to enhance energy, aid digestion, and treat skin ailments.

**Phytochemistry and Bioactivity:** High in vitamins C and E, minerals, and essential oils, this species has a rich antioxidant compound profile that lowers inflammatory responses in humans. Extracts have been shown to reduce anxiety, cortisol, and mental fatigue.

**Safety Precautions:** Do not use if pregnant or breastfeeding except under the supervision of a physician. If you take medications, consult a qualified health-care practitioner before taking schisandra; it has potential drug interactions.

## To Nicolai Vavilov

By the Shoots & Roots team

Nicolai Vavilov was an outstanding botanist who studied edible plants around the world and started the first seed bank. Tragically, he starved to death, imprisoned by Stalin. Many others starved protecting the seed bank during the 872-day Siege of Leningrad in World War II. The whole world benefits from Vavilov's seed bank, even today. We toast to you, our hero, Vavilov.

¾ oz (20 mL) Pine Nut Orgeat (see right)

1½ oz (45 mL) white rum

½ oz (15 mL) rum agricole

½ oz (15 mL) Schisandra Syrup (see page 125)

¾ oz (20 mL) lime juice

1 dash Citrus Grove Bitters (see page 69) or Grapefruit Peel Bitters (see page 51)

**Pine Nut Orgeat:** Toast 1 cup (135 g) pine nuts in a single layer in a skillet. Let cool, then blend the nuts in 1 cup (240 mL) water for at least 1 minute. Bring to a boil in a medium saucepan. Add 1 cup (200 g) sugar and 1 star anise and boil for another 2 minutes, stirring constantly. Turn off the heat and let cool for about 2 hours. Strain through a nut milk bag into an airtight storage container. Refigerate for up to 2 weeks. Yields about 1 cup (240 mL).

**Method:** Combine all the ingredients in a shaker and add ice. Shake and strain into a chilled 6-ounce (175-mL) glass. No garnish. Yields one 4 oz drink.

# Barley

**Latin Name:** *Hordeum vulgare* (L.)

**Botanical family:**
Poaceae (the True
Grass family)

**Common name(s):**
Barley

**Habitat:**
Barley is cultivated worldwide in temperate climates and montane areas of tropics.
In temperate areas barley is grown as a summer crop, and in tropical areas it is grown
as a winter crop.

**Parts used:**
Grain, whole plant for fodder

**Diversity:**
There are hundreds of modern varieties.

**Botany:** The plants are a predominantly self-pollinated, annual grass. Barley does not tolerate extreme cold so is rarely found above 5,000 feet (1,500 m). It is moderately tolerant of drought and salinity but not the cold.

**Cultural Uses:** Historically, barley has been used for bread, porridge, beverages, and fodder. Barley has been important for thatch, basketry, mudbrick, and pottery. Columbus's voyages brought barley to the New World, where cultivation was most successful in North America.

**Culinary Uses:** Half of the U.S. production of barley is for malting. The grain is also often used in soups and baby foods. Barley flour is made from ground pearl barley.

**Traditional Medicinal Uses:** In Argentina, Iran, and Italy, hot water extracts of the grains are taken to treat infections of the bladder and urinary tract. In Guatemala, Iran, Italy, and Korea, hot water extracts of the grains are used as an ointment to reduce inflammation.

**Phytochemistry and Bioactivity:** Dried stem extracts have shown antifungal activity. Germinated barley reduces colonic inflammation and improves symptoms of ulcerative colitis. Water extracts from the fermented root and dried grains have shown hypoglycemic activity.

**Safety precautions:** Several allergens have been isolated and characterized in barley. Barley flour and malt exposure can cause dermatitis and asthma. Barley can lead to symptoms of gluten allergy, such as gastrointestinal distress, dermatitis, hives, angioedema, and anaphylaxis.

# Crane Dance

By the Shoots & Roots Team

This is missing one element of a cocktail—a sweetener—but it does not need it. A good shochu is balanced and you want to taste the barley. The bitters rest on top to add to the aroma but do not adulterate the flavor of the shochu. Barley is the grain with the deepest history in beverage and food use. The genetic diversity in the barley crop, with lineages from Europe, north Africa, the Levant, and Asia, allows for it to be cultivated all over the world, and you can taste that diversity through the different varieties used in beverages.

2 oz (60 mL) all-barley shochu

5 drops Ki Japanese Arboreal Bitters (see page 67) or Alone in Kyoto Bitters (see page 53)

**Method:** Pour the shochu into a short glass over ice, drop the bitters on top, and enjoy.

# Baobab

**Latin Name:** *Adansonia digitata* (L.)

**Botanical family:**
Malvaceae (the Mallow family)

**Common name(s):**
Baobab, upside-down tree, dead rat tree, monkey bread tree, cream of tartar tree, Ethiopian sour gourd, Senegal calabash

**Habitat:**
Native to the African savanna. The tree has been introduced to many countries in Africa, as well as the Arabian Peninsula, the Malay Peninsula, and the Caribbean.

**Parts used:**
Fruit pulp, seed, leaves, flowers, roots, bark

**Diversity:**
There are nine species in the Adansonia genus and all are used by people.

**Botany:** Baobabs can grow to 80 feet (25 m) tall and have huge trunks 33 to 45 feet (10–14 m) in diameter. The root span can be wider than the tree is tall, helping it find water. Trunk bark is shiny and corklike, resembling elephant skin. Trees live for an estimated 1,500 years.

**Cultural Uses:** Baobabs have been traditionally valued as sources of food, water, shelter, and medicine. The bark and fruit shell are used as insect deterrents. The tree fiber makes excellent goods, from ropes to canoes. Tree tannins are used as a dye.

**Culinary Uses:** Nearly all the plant is edible. Leaves, fruit pulp, and seed are central ingredients in many dishes. Young roots are a common vegetable, and mature roots are cooked and eaten during famine. The wood can be chewed to extract water.

**Traditional Medicinal Uses:** The bark is used to fight pain. Leaves and flowers can be boiled and used to treat asthma, fatigue, kidney and bladder diseases, and digestive disorders. According to the United Nations, baobab was a common medicine for smallpox.

**Phytochemistry and Bioactivity:** Substances present in the stem and root bark have been attributed with strong antibacterial activity. Methanolic extracts of leaves and roots showed antiviral activity against herpes and sindbis virus.

**Safety Precautions:** Unrefined (aka unboiled or unhydrogenated) baobab oil contains antinutritionals known as cyclopropane fatty acids. Recent reports have warned that they can cause cancer. The oil is made from pressing the seed and is often marketed as a "cure-all."

## Bab's Dock

By Christian Schaal

Bab's Dock is an amazing getaway near Cotonou, Benin. To get there, you have to take a boat through dense mangroves and then reeds, which open to a paradise bar and lounge with tortoises and lizards roaming free. It was a most excellent break from roughing it during fieldwork. Baobab fruits contain a pulp that dries into powder, which is widely available in health food stores across the globe and is used as a common snack, beverage, or pudding across all communities of the African Savanna.

1½ oz (45 mL) Scotch

½ oz (15 mL) Amontillado sherry

½ oz (15 mL) lime juice

¾ oz (20 mL) Baobab Syrup (see page 125)

1 dash *Angostura Trifoliata* Bitters (see page 69) or Thai Market Bitters (see page 53)

Lime wheel and baobab powder (for garnish)

**Method:** In a shaker, combine all the ingredients and then add ice. Shake and fine strain into a coupe glass. Garnish with a lime wheel dusted with baobab powder on one side. Yields one 3½ oz drink.

# Ajwain

**Latin Name:** *Trachyspermum ammi* (L.) SPRAGUE. EX TURRILL

**Botanical family:**
Apiaceae (the Parsley family)

**Common name(s):**
Ajwain, ajowan, Ajowan caraway, bishop's weed, carom, yamini, yaminiki, yaviniki, jain, vamu, jevain, yom, ayanodakan

**Habitat:**
Native to Egypt and cultivated in India, Pakistan, Iran, Iraq, and Afghanistan. It is widely grown in arid and semiarid regions, where soil contains high levels of salts.

**Parts used:**
Schizocarp (dry fruit that splits up into sections—mericarps—when dry), roots, seed oil

**Diversity:**
Recent analysis demonstrated a population structure of three major groups.

**Botany:** Ajwain is a branched annual herb, 2 to 3 feet (60–90 cm) in height. Its stem is streaked with clusters of flowers known as umbels, each one containing up to 16 flowers. These are symmetrical, white, male, and bisexual, with bilobed petals. The fruit has two ovoid, ridged, grayish-brown mericarps, $1/16$ inch (2 mm) long and $5/8$ inch (1.7 mm) wide.

**Cultural Uses:** Ajwain is a highly valued, medicinally important seed spice and also used as a flavoring and condiment.

**Culinary Uses:** Thymol derived from the seed oil of ajwain is used in toothpaste and perfumery. The seeds are used in aphrodisiacs as well as to cure abdominal tumors, abdominal pains, and hemorrhoids. The pungent aromatic fruit is used as a flavoring in savory dishes, such as curries, breads, and pastry snacks.

**Traditional Medicinal Uses:** In traditional Ayurvedic medicine, ajwain is primarily used for stomach disorders (indigestion and flatulence). The seed oil is used for the treatment of gastrointestinal ailments, lack of appetite, and bronchial problems. Crushed fruit is used topically as a poultice.

**Phytochemistry and Bioactivity:** Studies suggest that ajwain has numerous medicinal properties, including carminative, antioxidant, antimicrobial, antinociceptive (blocking the sensation of pain), antifungal, antihypertensive, antispasmodic, broncho-dilating, and stimulating properties.

**Safety Precautions:** There are no safety precautions associated with ajwain. In fact, roasted ajwain is considered safe to be used by pregnant women.

# Bishop's Weed

### By Christian Schaal

There are many spice and food plants that connect Africa with India. One of them is ajwain, a delicious superpotent cumin-thyme grain. You'll see ajwain as an ingredient in lentil dishes throughout northern Africa, from Morocco to Ethiopia, and in many Ayurvedic recipes in India. We figured we could highlight this in a gin and tonic, drunk widely in India and Africa during the expansion of the British Empire. To get really historic with this cocktail, replace the gin with genever.

½ oz (15 mL) Ajwain-Infused Dry Vermouth (see right) or 1 dash Mediterranean Garden Bitters (see page 55)

1½ oz (45 mL) gin

¼ oz (7.5 mL) Simple Syrup (see page 124)

½ oz (15 mL) Meyer lemon juice

1-2 oz (30-60 mL) tonic

**Ajwain-Infused Dry Vermouth:** Toast a few teaspoons of ajwain fruit (which are schizocarps, a kind of fruit often mistakenly called seed) over medium heat in a dry skillet, stirring continuously so they don't burn. After about 5 minutes, or when the fruit has a roasted aroma, remove from the heat. Add ¾–1 tsp of the toasted ajwain fruit to 6 oz (175 mL) dry vermouth (you can make your own using the recipe in this book, or alternatively, we recommend Dolin brand dry vermouth). Blend for 30 seconds and let sit for 15 minutes. Strain. Store in a glass container in the refrigerator.

**Method:** In a shaker, combine the gin, vermouth, simple syrup, and lemon juice and shake with ice. Pour over fresh ice in a Collins glass. Top with tonic. Garnish with a thick Meyer lemon twist. Yields one 5 oz drink.

# Honeybush

**Latin Name:** *Cyclopia* spp.

**Botanical family:**
Fabaceae (the Pea family)

**Common name(s):**
Honeybush (Note: *Cyclopia* spp. are of a different genus than the Cape species used to make rooibos tea, which is also often referred to as honeybush tea)

**Habitat:**
*Cyclopia* spp. are endemic to the fynbos biome and occur on the coastal plains and mountainous regions of the Western and Eastern Cape provinces in South Africa.

**Parts used:**
Fermented and dried leaves and flowers

**Diversity:**
There are twenty-three species recognized in the genus *Cyclopia*.

**Botany:** The flowers have a circular depression in the base where the pedicel, or leaf stem, attaches. Honeybush tea plants have woody stems, a low leaf-to-stem ratio, and hard-shelled seed that need to be damaged naturally or artificially to germinate. Leaves vary in shape between species.

**Cultural Uses:** Traditionally, leafy shoots and flowers of the *Cyclopia* were fermented and dried to make a tea in the Cape region of South Africa. Honeybush tea from *Cyclopia* remained regional until the first branded product appeared in the 1960s, and a formal industry was established in the early 1990s. Powdered extracts from fermented honeybush are now used for cosmetics.

**Culinary uses:** The aromatic, caffeine-free tea can be processed fresh or fermented. Sold as a coarse blend of ground leaves, flowers, and stems, it blends well with fruit juices for iced tea. The flavor has been likened to spicy apricot preserves, with floral, honeylike, and dried fruit notes.

**Traditional Medicinal Uses:** Decoctions have been used traditionally in the Cape region as a restorative and expectorant. The tea is considered an appetite stimulant, often administered to children with stomach problems, for which it is well suited because of its lack of caffeine and tannins. It has been used to stimulate lactation, as a cure for colitis, and to alleviate arthritic pain.

**Phytochemistry and Bioactivity:** Substantial amounts of (+)-pinitol are present in honeybush tea, an expectorant with possible antidiabetic effects. The presence of flavonoids, xanthones, isoflavonoids, and coumestans have been documented in *Cyclopia* species, but all occur at different concentrations in the different species.

**Safety precautions:** None reported.

# Bettysbaai

### By the Shoots & Roots Team

*Cyclopia* (honeybushes) and *Pelargonium* (geraniums) are native to the Fynbos biome of South Africa. This is an area of high biodiversity, containing more than six thousand endemic species that are under serious threat due to climate change,. Honeybush is a culturally important tea in the Fynbos, used to perfume houses as it steeps. Combine fragrance with the relaxing effect of the skullcap in the bitters and you have a perfect cocktail for sitting back and gazing out at the sea.

1½ oz (45 mL) Strong Honeybush Tea (see right)

1¾ oz (50 mL) dark rum

½ oz (15 mL) honey-flavored liqueur or vodka

½ oz (15 mL) lemon juice

1 dash Demon Flower Bitters (see page 65) or Veracruz Bay Bitters (see page 56)

Geranium flowers (for garnish)

**Strong Honeybush Tea:** To make a concentrated tea, in a small saucepan, boil 1 heaping tablespoon of the dried honeybush leaves with 1 cup (240 mL) of water over medium heat for 15 minutes. Let cool and finely strain to remove the tiny leaves. Refrigerate for up to 1 week. Yields about ¾ cup (175 mL).

**Method:** In a shaker, combine all the ingredients and then add ice. Shake and strain into a rocks glass with a single large ice cube. Garnish with a sprig of *Pelargonium* (geranium) inflorescence that you can also use to stir the drink or with *Pelargonium* flowers frozen in an ice cube. Yields one 4½ oz drink.

# Pomegranate

**Latin Name:** *Punica granatum* L.

| **Botanical family:** | **Common name(s):** |
|---|---|
| Lythraceae | Pomegranate, fruit of the underworld, fruit of |
| (the loosestrife family) | paradise, seeded apple, anar, dadimah, dalimba |

**Habitat:**
Pomegranate is native to modern-day Iran and has been cultivated throughout the Mediterranean region and northern India.

| **Parts used:** | **Diversity:** |
|---|---|
| Arils, seed, fruit husk, leaves | More than five hundred pomegranate varieties have been described. |

**Botany:** The pomegranate grows as a deciduous shrub or small tree between 15 and 26 feet (5–8 m). Flaming orange-red flowers with crinkled petals and numerous stamen grow from slender, thorny branches. Leaves are glossy dark green. The brownish-yellow to purplish-red berries have a smooth, leathery skin.

**Cultural Uses:** In Persian mythology, pomegranates granted invincibility. In Greek mythology, they represented the permanence of marriage. In Judaism, each fruit has 613 seeds, one for each of the Bible's Commandments; in Christianity, it is linked with fertility. The Koran describes pomegranate trees in a heavenly paradise, and in Buddhism it is a blessed fruit, along with peach and citrus.

**Culinary Uses:** Fresh seed are sprinkled as a garnish for desserts and salads, while dried seed are used as a souring agent in many Indian dishes. The juice is a popular ingredient in wine and a range of dishes.

**Traditional Medicinal Uses:** Seed and juice have been consumed as a tonic for the heart, throat, and eyes, to treat bleeding noses and gums, for hemorrhoids, and for toning skin. The juice is consumed to reduce the effects of dental plaque. Seed extracts may slow the development of cataracts.

**Phytochemistry and Bioactivity:** The edible arils and seed provide vitamin C, vitamin K, folate, and dietary fiber. Pomegranates have been shown to have two to three times the antioxidant capacity of red wine or green tea.

**Safety Precautions:** While pomegranate juice is safe during pregnancy, pregnant women should avoid extracts made from pomegranate rind, because it can induce contractions and early labor.

# Blood on the Moon

By Christian Schaal

As Turkey stretches from West to East, bridging Europe and Asia, the landscape is dotted with pomegranate trees. Their diversity can be tasted, sliced, and juiced like citrus at roadside stands. South of Istanbul, the landscape becomes barren; salt and mineral deposits ooze from the soil and give way to a lunar surface of Cappadocian fairy chimneys.

1 oz (30 mL) bourbon

1 oz (30 mL) mezcal

1 oz (30 mL) fresh pomegranate juice

½ oz (15 mL) chili pepper liqueur

½ oz (15 mL) lime juice

½ oz (15 mL) Simple Syrup (see page 124)

1 dash Hanging Gardens of Babylon Bitters (see page 66) or Biennial Bitters (see page 54)

Lime wheel and pomegranate seeds, or ancho chili powder, salt, and sugar (for garnish)

**Method:** In a shaker, combine all the ingredients and shake with ice. Pour over fresh ice in a rocks glass. Garnish with a lime wheel, pomegranate seed, or with a rim coated in ancho chili powder, salt, and sugar. Yields one 5 oz drink.

# Saffron

**Latin Name:** *Crocus sativus* (L.)

**Botanical family:**
Iridaceae (the Iris family)

**Common name(s):**
Saffron crocus, autumn crocus

**Habitat:**
A perennial plant that is cultivated from the western Mediterranean to India, Tibet, and China. Recently, it has been introduced into cultivation in Australia, Mexico, Argentina, and New Zealand.

**Parts used:**
Stigmas

**Diversity:**
The genus *Crocus* includes eighty-five to one hundred species.

**Botany:** Plants grow from fall to late spring and survive dry summers by means of a compact corm, a bulb-like underground stem. The plants grow five to eleven leaves. Flowers are fragrant, deep lilac-purple with darker veins and a violet stain in the throat.

**Cultural Uses:** Frescoes of 1600 BCE Crete portray the flowers picked by monkeys and girls; others show goddesses using saffron in a remedy and to treat a bleeding foot. In ancient Persia, saffron perfumed the dead; in ancient Rome, it was worn as mascara; and in ancient Egypt, Cleopatra allegedly used it in baths of milk to color her skin.

**Culinary Uses:** Saffron is used as a flavoring and coloring agent for breads, soups, sauces, rice, and desserts. It plays an important role in many traditional regional cuisines, including paella, bouillabaisse, and risotto, and in Milanese and Persian cuisine. The aroma has been described as like hay and "sea air."

**Traditional Medicinal Uses:** Documented medical uses of saffron include the treatment of eye problems, gastrointestinal ailments, a tonic, and an antidepressant. In Hellenistic Egypt, an emulsion of immature saffron flowers mixed with roasted beans was used to treat urinary tract conditions.

**Phytochemistry and Bioactivity:** About fifty constituents of saffron have been identified. Certain extracts of stigmas and petals showed anti-inflammatory effects; stigma extract also showed some activity against acute inflammation.

**Safety Precautions:** Be careful to avoid consuming saffron in high dosages (more than 200 mg) on a chronic basis (more than 2 months) to prevent nausea, vomiting, diarrhea, and bleeding. As with all herbs and spices, pregnant women should consume saffron in moderation.

# Superbloom

By the Shoots & Roots Team
and Christian Schaal

Surprisingly, while saffron is incredible in meals and desserts, it's hard to strike the right balance in cocktails. Usually it's either undetectable or it's overpowering. With this beer cocktail, we think we hit the jackpot. Saffron is worth more than its weight in gold, and it is the most adulterated spice in the world. It is made from crocus stigmas, commonly called "threads"; real saffron spice should never be in clumps of more than two stigmas, and the color it dyes water is yellow. Imposter species will dye water red or brown and won't make your drink taste of bitter honey and heady perfume.

---

1 oz (30 mL) gin

½ oz (15 mL) lime juice

¼ oz (7.5 mL) peach liqueur

½ oz (15 mL) Saffron Syrup
(see page 126)

1 dash Danube Bitters
(see page 52)

10 oz (300 mL) India Pale Ale
beer

**Method:** Build in a 12-oz (355-mL) beer glass. Add gin, lime juice, peach liqueur, Saffron Syrup, and Danube Bitters, stir to mix briefly, and fill the glass slowly with IPA. Give it a single additional stir and serve. Yields one 10 oz drink.

# Caraway

**Latin Name:** *Carum carvi* (L.)

**Botanical family:**
Apiaceae (the Parsley family)

**Common name(s):**
Caraway, Persian cumin, carvies

**Habitat:**
The annual form is native to the eastern Mediterranean and north Africa; the biennial form is native to central Europe, from the Alps to the Caucasus. Wild caraway stretches from Siberia to India, Africa, and northern Europe.

**Parts used:**
Fruit (usually called seed), leaves, roots

**Diversity:**
Caraway diversity is poorly documented, although differences between wild and cultivated types have been described.

**Botany:** These plants are about 1 foot 8 inches (50 cm) tall and have feathery leaves. They have white or pink clusters of flowers that bloom annually or biennially. If grown in a greenhouse, it needs a wind ventilator, otherwise the fruit, tiny achenes, will be devoid of seed.

**Cultural Uses:** It is used in food, beverages, medicines, cosmetics, soaps, mouthwashes, and skin rubs. The plant has a rich folklore. Superstitions hold that an object containing caraway cannot be stolen, and that it keeps lovers interested in each other. In north and central Europe, caraway seeds are used in confectionary celebrating the fall equinox.

**Culinary Uses:** Seed are used as a popular flavoring in liqueurs, savory cooking, and desserts. Tender leaves can be boiled to flavor soup. Often used in sauerkraut and other cabbage products. Caraway is a component of harissa, a spicy condiment.

**Traditional Medicinal Uses:** Caraway is an important part of European, Indian, Turkish, Persian, Russian, and Middle Eastern traditional therapies. Considered a carminative and antispasmodic astringent, it treats various digestive disorders, morning sickness, and colic, and it is said to improve liver function. Vapors are used to relieve lung trouble. Caraway water is used in pediatric medicine, and as an alcohol-and-oil mixture, it is used to treat scabies.

**Phytochemistry and Bioactivity:** Caraway fruit is high in essential oils and oleoresins, that is, naturally occurring mixtures of oils and resin. Both fruit and root extracts have high free radical scavenging activity, which may underlie the anticancer, antistress, and antidiabetic properties. The essential oils inhibit growth of fungus and bacteria.

**Safety Precautions:** Caraway can cause vomiting and diarrhea in dogs and horses.

# Delicatessen

By Christian Schaal

Rye carries a rich story in German folklore, in which the *Kornmutter* field spirit causes the grains swaying in the wind to turn an auspicious black, and where runaway pigs give birth to rye to explain the origin of this Eastern European grain to Britain. Rye and caraway are not only what define sweet and earthy rye bread, but are the sustenance of the worker and the immigrant. Like the people that relied on them, they grew to be the proud symbol of a healthy, wholesome society.

---

1½ oz (45 mL) Caraway-Infused Rye Whiskey (see right)

½ oz (15 mL) bourbon

Bar spoon aquavit

Bar spoon simple syrup

1 dash *Angostura Trifoliata* Bitters (see page 69) and 2 dashes Raisin in the Sun Bitters (see page 68), or 3 dashes Turkish Coffee Bitters (see page 50)

Lemon peel (for garnish)

**Caraway-Infused Rye Whiskey:** Add 1 tsp caraway fruit (which are schizocarps, each containing a single seed) to 6 oz (175 mL) rye whiskey. Blend the mix for 30 seconds and let rest in the blender or another container for 3 hours; then finely strain. Yields about ¾ cup (175 mL) for about four drinks.

**Method:** Add all the ingredients except the bitters to a mixing glass, then ice and stir until chilled. Meanwhile, chill a rocks glass. Rinse the glass with a few dashes of bitters (roll the bitters inside the glass to coat it and let the excess pour out), strain the drink into the chilled rocks glass, add a lemon peel, and you're done. Yields one 3 oz drink.

# Grape

---

**Latin Name:** *Vitis vinifera* (L.)

| **Botanical family:** | **Common name(s):** |
|---|---|
| Vitaceae (the Grape family) | Grape, grapevine, table grape, wine grape, wild grape |

**Habitat:**
Grapes are native to western and middle Asia, the Caucasus, northern Africa, and parts of Europe. Native habitats are alluvial and humid forests, and woody vegetation along the sides of rivers and streams.

| **Parts used:** | **Diversity:** |
|---|---|
| Fruit, leaves and young stems, sap | There are roughly 10,000 varieties of grape, and nearly a dozen other species used as rootstocks. |

---

**Botany:** Grapes are perennial lianas, or long-stemmed woody vines, growing to 105 feet (32 m) long. Grapes have alternate, long lobed leaves 2 to 8 inches (5–20 cm) long and wide. The tiny, yellow-green flowers develop into green, red, purple, blue, or black grapes.

**Cultural Uses:** Archaeological evidence suggests that they were cultivated in modern-day Iran as early as 3,500 BCE, and spread out from there. Ancient societies of Egypt, Crete, Mesopotamia, Greece, and Rome advanced the art and science of viticulture.

**Culinary Uses:** Grapes have a long history of use for food, beverage, and medicine, mostly eaten fresh and as the basis of most fermented wine.

**Traditional Medicinal Uses:** In traditional Chinese medicine, grape fruit are used to strengthen the body and remedy debilitations, especially in the debilitated. In Ayurveda, grapes are used to treat the nervous, digestive, circulatory, respiratory, and reproductive systems. Native Americans used various parts of wild grape for the treatment of diabetes and snakebites. Grape seed extract is used for treating tooth decay, eye diseases, blood circulation, and digestive disorders.

**Phytochemistry and Bioactivity:** Grapes are rich in flavonoids, minerals, and vitamins; their overall phytochemical profile has liver protective effects and antimicrobial, antihyperlipidemic, anti-inflammatory, antidiabetic, antioxidant, and anticancerous activities.

**Safety Precautions:** Grapes are not associated with toxicity and are regarded as safe for pregnant women. However, pregnant women should avoid grape seed extract, because many commercial extracts contain synthetic substances.

## House of Panda

By Kevin Denton

Here's a drink paying tribute to the close relationship between plants and microbes. It's a sherry drink, and if you haven't looked into sherry or the fungal "flor" native to Andalusian grapes that develops this wine's flavor, you're in for a treat. This grape-base vermouth boasts Asian ingredients, such as ginger and clove, but also numerous native *Artemisia* species (related to wormwood and mugwort).

1½ oz (45 mL) Manzanilla sherry

1½ oz (45 mL) ginger-infused vermouth (You can make our Amphora Dry Vermouth—see page 58—with additional ginger)

½ oz (15 mL) reposado tequila

1½ oz (45 mL) soda

2 dashes Ki Japanese Arboreal Bitters (see page 67) or Thai Market Bitters (see page 53)

Grapefruit or lemon peel (for garnish)

**Method:** Build the cocktail by adding all the ingredients except the bitters in a Collins glass with large cubes of ice. Top with soda and a heavy dose of Ki Japanese Arboreal bitters. Garnish with a large grapefruit or lemon peel. Yields one 5 oz drink.

# Cacao

**Latin Name:** *Theobroma cacao* (L.)

**Botanical family:**
Malvaceae (the Mallow family)

**Common name(s):**
Chocolate, cacao, cocoa

**Habitat:**
Native to the upper Amazon with centers of diversity in Ecuador, Bolivia, and Peru, cacao was domesticated by the Mayans. Semiwild strands are found throughout Bolivia and Southern Mexico.

**Parts used:**
Beans and pulp are used in the production of cocoa and cocoa butter

**Diversity:**
There are twenty-four species of *Theobroma*.

**Botany:** The cacao tree is a shade-tolerant understory species. The fruit is distinct; the pods grow directly from the main stem and branches. Pods range widely from small and round to long, warty, and deeply grooved, and they vary from green to red at maturity, depending on genetic background and growing environment.

**Cultural Uses:** Beans were not only food and drink for the Maya in the Yucatán but also their currency. A Nahuatl document in 1545 describes turkey hens and rabbits to be worth 100 full cacao beans (or 120 shrunken). Cacao symbolized blood to the Aztecs and was central in the ritual ceremonies for new Jaguar Knights, the bravest Aztec warriors.

**Culinary Uses:** Cocoa is separated from cocoa butter to form a dark chocolate liquor, then processed into powder and recombined with cocoa butter in different proportions to form different qualities of chocolate.

**Traditional Medicinal Uses:** The Badianus Manuscript of Mexico (1552) provides a critical understanding of the Mesoamerican use of chocolate for nutrition and healing, including for angina, constipation, dental problems, dysentery, dyspepsia, indigestion, fatigue, gout, and hemorrhoids.

**Phytochemistry and Bioactivity:** Chocolate has a rich flavanol profile that has been demonstrated to support cardiovascular health. There is strong evidence for an association of dark chocolate consumption with increased flow-mediated vasodilatation, reducing blood pressure. Some parts of chocolate could also affect the release of certain neurotransmitters in the brain, which lessen anxiety and trigger production of endorphins, the "feel-good" hormones.

**Safety Precautions:** Those with allergies to caffeine and theobromine should avoid chocolate.

## Chocolate Jesus

By Christian Schaal

Linnaeus immortalized cacao as food for the gods, and many have considered it more valuable than gold. Montezuma drank his chocolate from single-use golden goblets, and Tom Waits prefers his Chocolate Jesus to a cup of gold. This cocktail could potentially be a dessert beverage, as cacao often is. *Amburana* is a genus of trees in the legume family, native to Brazil. The wood has a distinct vanilla bouquet. The two marry wonderfully in this short drink—a toast to chocoholics throughout history.

1½ oz (45 mL) Cacao
Bean-Infused Cachaça
(see right)

½ oz (15 mL) Amphora Dry
Vermouth (see page 58) or
1 dash Xocolatl Bitters
(see page 50)

½ oz (15 mL) rhubarb amaro
(we use Zucca Amaro)

**Cacao Bean-Infused Cachaça:** Mix 2 cups (230 g) of unsweetened cacao nibs (the seed) with one 25-oz (750-mL) bottle Avua Amburana and let infuse for 24 hours at room temperature. We just use plastic quart containers for this step. Strain.

**Method:** In a mixing glass, combine all the ingredients and then add ice. Stir and strain into a short 8-oz (175-mL) rocks glass over a large ice cube. No garnish. Yields one 3 oz drink.

# Habanero Chili Pepper

**Latin Name:** *Capsicum chinense* JACQ.

**Botanical family:**
Solanaceae
(the Nightshade family)

**Common name(s):**
Habanero, bonnet pepper, yellow lantern chili,
seven-pot chili, Datil, scorpion pepper

**Habitat:**
Hot, humid climates of South and Central America. Now grown widely in tropical
and subtropical areas of the globe, such as in southwest China.

**Parts used:**
Fruit, leaves

**Diversity:**
The genus itself, *Capsicum*, is small, containing
just over twenty species.

**Botany:** The species is a fast-growing evergreen shrub that can grow up to 5 feet (1.5 m) tall, but habanero plants are usually small, about 1 foot 8 inches (50 cm). It has subtle white self-fertile flowers with five to six petals, and berries of various sizes. Pungent compounds are concentrated around the seeds.

**Cultural Uses:** South Americans probably grew and ate chili before the earliest archaeological evidence, which dates back nearly five thousand years. Chili peppers have been important medicines treating many parts of the body through their effects on nervous, cardiovascular, and digestive systems.

**Culinary Uses:** The earliest evidence suggests *C. chinense* was prepared with potato, corn, and other starches. It is still used as a condiment for atole, a traditional hot corn-and-masa beverage, in the Mayan Yucatán.

**Traditional Medicinal Uses:** The fruit have been used as a tonic and antiseptic to stimulate digestive systems and circulation and to increase perspiration. Applied topically, it increases blood supply, reducing pain sensitivity. It is used to treat asthma and the cold stage of fevers and for various digestive problems.

**Phytochemistry and Bioactivity:** The specific type of pain provided by capsaicin is mimicked by only certain tarantula venoms. These compounds have strong antioxidant activity and influence fat metabolism, playing an important role in high fat and high carbohydrate diets. The fruit extract has antiseptic properties.

**Safety Precautions:** Consuming large quantities can cause difficulty in breathing and convulsions. Chili pepper extract is used in weaponry as a spray, so avoid breathing vapors or getting extract in your eyes.

# 92° in the Shade

### By Christian Schaal

Mexico generally gets credit for chili pepper origins, but some species of domesticated *Capsicum* came from the Amazon, namely the exceedingly pungent varieties of habanero. The name itself, habanero, and the lack of a Mayan name suggest it arrived via Cuba.

---

1½ oz (45 mL) Habanero Cachaça (see right)

1 oz (30 mL) unsweetened coconut milk

¾ oz (20 mL) pineapple juice

½ oz (15 mL) lime juice

½ oz (15 mL) Simple Syrup (see page 124)

1 dash Raisin in the Sun Bitters (see page 68) or Chicha Morada Bitters (see page 54)

Pineapple (for garnish)

**Habanero Cachaça:** Determine how much you want to make, given 1½ oz are all that's needed per drink. With a spoon or pestle, press to muddle (mash) ¼ habanero, including seeds, for every 5 oz (150 mL) cachaça in a metal shaker cup. Let sit for 30 seconds to 2 minutes, depending on desired pungency (think of 2 minutes as five-stars spicy).

**Method:** In a shaker, combine all the ingredients except the bitters and then add ice. Shake and strain into a coupe glass and garnish with a pineapple wedge and a dash of Raisin in the Sun Bitters. Yields one 4½ oz drink.

# Lúcuma

**Latin Name:** *Pouteria lucuma* (RUIZ AND PAV.) (KUNTZE)

**Botanical family:**
Sapotaceae
(the Spondilla family)

**Common name(s):**
Lúcuma, lucmo, tiesa, eggfruit, cumala, rucma, marco, mammon

**Habitat:**
Lúcuma is native to temperate elevations of Andean valleys and the highlands of Peru, Ecuador, and Chile. It grows at elevations of 8,800 to 9,800 feet (2,700–3,000 m).

**Parts used:**
Pulp of the fruit

**Diversity:**
Today, the most common cultivars are Seda and Palo; the former is mainly consumed fresh while the latter is primarily used to make ice cream.

**Botany:** Lúcuma is an evergreen tree that grows to 26 to 50 feet (8–15 m) tall with a dense crown. Its branches produce a white latex. The fruit has an oval shape and green to yellowish skin when ripe. Unripe fruit contains latex. The seeds are glossy black; often there are two seeds in the fruit but there can be up to five.

**Cultural Uses:** By 2000 BCE, lúcuma was moderately abundant at archaeological sites in Huaynuna and El Paraíso. Lúcuma fruit have also been found depicted as art on ceramic vessels at burial sites of the indigenous people of coastal Peru, including the Moche people. It is mostly consumed in Peru and Chile. There are two popular desserts that used lúcuma in Chile: *merengue con salsa de lúcuma* and *manjar con lúcuma*.

**Culinary Uses:** The flesh of the fruit can be eaten raw, but it is often processed as a flavoring for ice cream, bakery products,

yogurt, milkshakes, and juice. The fruit pulp is often frozen or made into a flour for these processed foods and is used for preserves. Lúcuma has a flavor with similar notes to sweet potato, maple syrup, and butterscotch.

**Traditional Medicinal Uses:** Lúcuma has been traditionally cultivated in Andean and coastal communities in Peru for nutrition, fertility, and longevity. It provides a notable source of plant protein and, despite its sweet taste, has a low sugar concentration.

**Phytochemistry and Bioactivity:** Lúcuma fruit pulp is rich in beta-carotene, fiber, vitamins (especially vitamin B3), and iron. It has been shown to have relatively high antioxidant activity and α-glucosidase inhibitory activity (useful for diabetes management).

**Safety Precautions:** Lucuma is not associated with safety precautions.

# Ica Shuffle

### By Christian Schaal

Lucuma is a fruit native to Peru. The fresh fruit is similar to egg yolk, but the powdered extract has a delicious sweet raisin flavor. Damiana is a mood-enhancing subtropical herb in the passionflower family that has a cooling figlike flavor. Damiana tends to be bitter when steeped in water or alcohol for a long time. We combined these two South American plants, and added yogurt to balance out the flavors and get as much damiana in a drink as possible—it's a prized aphrodisiac for both men and women.

1½ oz (45 mL) Damiana-Infused Pisco (see right)

¾ oz (20 mL) Lucuma Simple Syrup (see page 124)

½ oz (15 mL) lime juice

1 tsp Greek yogurt

1 tsp crème de cacao

1 dash Xocolatl Bitters (see page 50)

Sprig of lavender (for garnish)

**Damiana Leaf-Infused Pisco:** Blend ½ tsp dried crushed damiana leaves with 6 oz (175 mL) pisco, let rest for 15 minutes, and strain.

**Method:** In a shaker, combine all the ingredients and shake with ice. Double strain into a stemless wine glass with fresh ice. Garnish with a sprig of lavender. Yields one 3 oz drink.

# Allspice

| **Latin Name:** *Pimenta dioica* (L.) MERR. | |
|---|---|
| **Botanical family:**<br>Myrtaceae (the Myrtle family) | **Common name(s):**<br>Allspice, pimento, Jamaica pepper, Clove pepper, Eugenia piment, pimienta inglesa, Tabasco pepper |
| **Habitat:**<br>Widespread in the West Indies and South America, production of allspice is concentrated in Jamaica. Although allspice has both male and female parts, it requires cross-pollination and has not been successful outside the American tropics. | |
| **Parts used:**<br>Dried, unripe berry | **Diversity:**<br>There is limited evidence describing the diversity of this species. |

**Botany:** The species epithet *dioica* described a peculiar feature of the plant. Although the flowers are bisexual, some trees are predominantly functionally male and others are functionally female, with little viable pollen but heavy crops of fruit.

**Cultural Uses:** In sea voyages from the seventeenth to nineteenth centuries, allspice berries were often used to preserve meat, and they continue to be used today in Scandinavia as a fish preservative. Topically, it is applied for muscle pains, toothaches, and as an antiseptic.

**Culinary Uses:** The fruit is picked green, sun dried, and sold whole or powdered. The spice has a complex taste similar to a mix of cinnamon, juniper, clove, and nutmeg.

**Traditional Medicinal Uses:** Historically, allspice has been ingested to treat intestinal gas and indigestion, stomachaches, heavy menstrual bleeding, vomiting, diarrhea, fever, flu, and colds. Allspice may have immune-stimulating effects.

**Phytochemistry and Bioactivity:** Allspice contains eugenol, used in dentistry as an antiseptic. Analgesic effects have been noted, but the mechanism of action is not well described. Allspice is rich in antioxidant compounds, the oil demonstrates strong insecticidal activity against termites, and powdered allspice shows antifungal activity.

**Safety Precautions:** There are known allergies and hypersensitivities to allspice, particularly contact dermatitis in people regularly exposed to allspice in food-service jobs. Allspice in large amounts might interact negatively with anticoagulant medicines and increase the risk of bleeding in the presence of eugenol; it is reported to have antiplatelet activity, but the evidence for this is limited.

## Dia Siete

By Christian Schaal

If you don't mind some evening caffeine, this is a great dessert shot. Dried unripe allspice berries are a prized spice from Central America. The fruit resembles cinnamon, nutmeg, and clove combined and is essential in Caribbean jerk seasoning. It was brought to the Old World in the sixteenth century and now has a name in more than fifty languages. With this drink, we bring this plant back to Southern Mexico, where most cacao, coffee, and various agaves for mezcal and tequila are also grown.

1½ oz (45 mL) Allspice-Infused Blanco Tequila (see right)

1 shot of espresso

½ oz (15 mL) mezcal

½ oz (15 mL) crème de cacao

3 to 5 drops Demon Flower Bitters (see page 64)

**Allspice-Infused Blanco Tequila:** Combine 1 tsp allspice fruit with 6 oz (175 mL) tequila blanco. Blend for 1 minute and let rest for 30 minutes. Strain.

**Method:** Pour on espresso shot (about 1 oz/25 mL) and let it come to room temperature. In a shaker, combine all the ingredients except for the bitters, and then add ice. Shake and pour into a tall shot glass. Drop bitters on the froth. No garnish. Makes two tall (2-oz/60-mL) shots. Yields two 2 oz drinks.

# Vanilla

---

**Latin Name:** *Vanilla* spp.

**Botanical family:**
Orchidaceae
(the Orchid family)

**Common name(s):**
Vanilla, bourbon vanilla (the name derives from *vaina*, a Spanish word for "sheath")

**Habitat:**
Vanilla plants originated in southeast Mexico and Guatemala and have now been intensively spread across other regions of the world.

**Parts used:**
Pods (commonly referred to as beans), seed

**Diversity:**
The *Vanilla* genus contains approximately 110 species.

---

**Botany:** Vanilla orchids are vines that climb trees using thick roots; they do not root in soil. The flowers are large, greenish yellow, and can self-pollinate to produce fruit (pods). However, they require a pollinator to transfer the pollen sac to the stigma on the single day the flower opens. Bees and bats are the seed dispersers.

**Cultural Uses:** Vanilla is widely used in food, beverages, perfume, incense, and cigars. Green pods were used to treat snakebites and other wounds. The folklore of the Totonac tells that vanilla first grew from where the lover of a princess was slain.

**Culinary Uses:** Vanilla pods, often called beans, are used for flavoring. The bourbon type grown in Madagascar and Reunion has the most intense flavor. Mexican vanilla is softer and fresher in aroma. Tahitian vanilla has the most floral flavor.

**Traditional Medicinal Uses:** Scholars occupying Mexico during the time of New Spain noted that vanilla-water decoction was used to help promote urination, to strengthen digestion, diminish flatulence, heal female ailments, and as a brain stimulant.

**Phytochemistry and Bioactivity:** Vanilla fermented fruit contains about 2 percent vanillin. Vanillin protects chromosone and DNA integrity, and it has been shown to have repairing properties. Vanillin is also anticarcinogenic and antimicrobial.

**Safety Precautions:** Vanilla is not known to be toxic, but some warnings have indicated possible skin contact irritation in some individuals as well as insomnia and headache for people working in vanilla processing.

# Mantis

By Christian Schaal

Besides fermented fruit pods of the prized vanilla orchid, the red color and the grapefruit tell Central American stories. Campari and Cappelletti, both bright red herbal liquors, used to be colored red thanks to the carmine pigment from ground-up cochineal beetles.

---

1½ oz (45 mL) bourbon

¾ oz (20 mL) sweet vermouth

1 drop natural vanilla extract

½ oz (15 mL) Cappelletti Aperitivo Americano Rosso

¼ oz (7.5 mL) Campari

1 dash Mexican Molcajete Bitters (see page 66) or Turkish Coffee Bitters (see page 50)

Grapefruit (for garnish)

**Method:** In a mixing glass, combine all the ingredients, add ice, and stir for about a minute until chilled (when about 10 percent has melted). Pour into a rocks glass with a large ice cube. Garnish with a grapefruit twist. Yields one 3 oz drink.

# Prickly Pear

**Latin Name:** *Opuntia ficus-indica* (L.) MILL.

**Botanical family:**
Cactaceae (the Cactus family)

**Common name(s):**
Smooth prickly pear, Indian fig, mission cactus, tuna, nopal, Barbary fig

**Habitat:**
Originally thought to be native to south-central Mexico, the prickly pear was domesticated in pre-Columbian times and widely introduced elsewhere, therefore, its precise native range is unknown. It has spread invasively in the wild.

**Parts used:**
Stems, fruit

**Diversity:**
Genotypes cultivated in Mexico show high levels of diversity.

**Botany:** Prickly pear cacti are perennial shrubs growing 10 to 16 feet (3–5 m) with thick succulent and oblong stems called cladodes. Mature cladodes produce flowers that range from green to deep red in color. The *Opuntia* species range from diploid (two sets of chromosomes) to octaploid.

**Cultural Uses:** Opuntia cacti (stem and fruit) have long been an important source of nutrition in the Americas, particularly for people and livestock in arid regions during drought periods. The plant bears symbolic importance as the national plant of Mexico and is depicted on the coat of arms on the Mexican flag.

**Culinary Uses:** In Mexico, cactus pads (*nopales*) are served sliced, similar to green beans. The fruits are eaten fresh and also canned. It is increasingly common as a flavoring for beverages and an ingredient in cocktails and blended beverages.

**Traditional Medicinal Uses:** Traditionally, stems have been used to treat diabetes in Mexico. The fruit, cladodes, and flower infusions have been used to treat ulcers, allergies, fatigue, and rheumatism and as a diuretic. Spanish conquerors brought the vitamin C-rich fruit on voyages as a remedy for scurvy.

**Phytochemistry and Bioactivity:** Several antioxidant compounds have been detected in the fruit and stems. Beta-sitosterol has been identified as the active anti-inflammatory agent from stem extracts. Seed oil and freeze-dried cladodes have demonstrated an ability to lower blood LDL-cholesterol by 10 to 15 percent.

**Safety Precautions:** Be careful to avoid the spines, as well as the areoles, which are covered with small, deciduous barbed spines called glochidium or glochids that can be especially irritating.

# Bed of Spines

By Christian Schaal

*Opuntia* is an iconic cactus found, surprisingly, all over the world. It is hard to find in stores, but you might have better luck finding it growing wild. All *Opuntia* fruit is nontoxic. If you can't find a chili liquor, experiment with your own chili infusion in rum or cachaça.

---

¼ oz (7.5 mL) Fresh Prickly Pear Juice (see right)

2 oz (60 mL) bourbon

½ oz (15 mL) chili liqueur

½ oz (15 mL) lime juice

¼ oz (7.5 mL) Simple Syrup (see page 124)

Light pinch smoked sea salt

1 dash Mexican Molcajete Bitters (see page 66) or Biennial Bitters (see page 54)

Prickly pear slices or lime wheels (for garnish)

**Fresh Prickly Pear Juice:** Remove the spines from the prickly pears, then cut off the skin and push them through a juicer or a fine mesh metal strainer to create a thick juice. If you aren't getting much yield from your pears, put the peeled pears into a blender with about the same volume of water, blend, and strain.

**Method:** In a shaker, combine all the ingredients and then add ice. Shake and strain into a clear glass of choice that has ice and prickly pear slices or lime wheels. Yields one 4 oz drink.

# Avocado

**Latin Name:** *Persea americana* MILL.

**Botanical family:**
Lauraceae (the Laurel family)

**Common name(s):**
Avocado, alligator pear, avocado pear, butter fruit, ahuácatl

**Habitat:**
Native to south-central Mexico, it occurs throughout South and Central America and is cultivated in tropical and Mediterranean climates.

**Parts used:**
All parts, especially fruit and leaves

**Diversity:**
There are approximately eighty-one species of *Persea*.

**Botany:** The avocado is an evergreen tree growing to 70 feet (21 m). The leaves are smooth and leathery at maturity and spirally arranged on the stem. Small white flowers are wind or insect pollinated. The fruit takes months to develop to full size on the tree and are picked to ripen off the tree.

**Cultural Uses:** The Nahuatl name *ahuácatl* translates to "testicle." All plant parts are used. The Aztecs used them as a sexual stimulant. Folk recipes for pest poisons use the avocado pits. Throughout Mesoamerica, the Inca, Olmec, and Maya developed a wide knowledge of and innovative uses for the trees. The fourteenth month in the Mayan calendar is named for the avocado.

**Culinary Uses:** For the Maya, uses range from fruit as baby food to being a worm repellant and antidiarrheal; impressively, they had obtained other species of avocados from elsewhere by 3400 BCE.

The Valdivia culture of western Ecuador spread the cultivation of the avocado plant.

**Traditional Medicinal Uses:** The leaves and bark are used to induce menstruation and abortion. Leaf tea is used to improve kidney function and treat colds. In Chilean traditional medicine, the leaves are used to treat respiratory problems and as a stomach tonic to relieve diarrhea, gas, and bloating.

**Phytochemistry and Bioactivity:** Avocados are high in potassium, and vitamins A and E. The more mature fruit is higher in oleic acid, which helps to control diabetes. Avocado applied topically helps protect skin from ultraviolet (UV) damage.

**Safety Precautions:** Avocado leaf and bark extracts should be avoided during pregnancy. All of the plant, except for the fruit pulp, is toxic to livestock and fish, causing edema and cardiomyopathy.

# Americani

By the Shoots & Roots Team

Avocado leaves, known as *hoja de aguacate*, are a common anise-scented food flavoring as well as medicine. With the avocado cultivation so popular in the subtropics, we see a lot of innovation in uses of the plant parts. We found nuns making special vermouth with avocado leaves in rural Togo and just had to bring a bottle back.

1 oz (30 mL) Avocado Leaf-Infused Dry vermouth (see right)use regular dry vermouth with an addition of 1 dash of Veracruz Bay Bitters (see page 56)

1½ oz (45 mL) reposado te quila

1 tsp Honey Syrup (see page 124)

¼ oz (7.5 mL) lemon juice

Avocado skin or leaf (for garnish)

**Avocado Leaf-Infused Dry Vermouth:** Blend two dried avocado leaves for every 6 oz (175 mL) vermouth (you can make the Amphora dry vermouth on page 58 or use any dry vermouth) for 30 seconds, let rest for 1 week in an airtight container in the refrigerator, then strain.

**Method:** In a mixing glass, combine all the ingredients and then add ice. Stir and strain into a chilled coupe glass. Garnish with an avocado skin or float a piece of avocado leaf.

# Blueberry

**Latin Name:** *Vaccinium corymbosum* (L.) and other *Vaccinium* spp.

| **Botanical family:** | **Common name(s):** |
|---|---|
| Ericaceae (the Heath family) | Blueberry, highbush blueberry, American blueberry, rabbiteye blueberry, mayberry, whortleberry |

**Habitat:**
Blueberries are widespread throughout temperate areas of the world, but they are generally restricted to acidic soil.

| **Parts used:** | **Diversity:** |
|---|---|
| Fruit | There are approximately 450 species of *Vaccinium*. |

**Botany:** Highbush blueberry is an upright, crown-forming shrub that grows 6 to 12 feet (1.8–3.6 m) tall. The twigs are yellow-green to reddish in winter, covered with wartlike dots. The deciduous leaves are slightly waxy with hairs on the veins beneath. The small, urn-shape flowers are white to tinged with pink. The fruit ripens from April to October.

**Cultural Uses:** Blueberries are popular in ornamental landscaping and provide an important summer food source for birds and mammals. The indigenous peoples of North America referred to them as "star berries."

**Culinary Uses:** The berries are eaten raw, smoke or sun dried, boiled, and baked in a wide range of confections. Native Americans pounded the berries into meat to flavor and help preserve it. Other edible *Vaccinium* spp. include cranberries, bilberries, lingonberries, and huckleberries.

**Traditional Medicinal Uses:** Blueberry leaf tea was used to treat urinary tract infections, diabetes, and boost appetites. Juice has been used for sore throats and inflamed gums and as a cough remedy, while leaf and bark decoctions have been used for oral sores, as an antidiarrheal, and an antiseptic. Women have used blueberry for labor pains and as a tonic after miscarriage.

**Phytochemistry and Bioactivity:** Blueberries have one of the highest concentrations of iron of any temperate fruits.

**Safety Precautions:** Excessive consumption of blueberry leaf tea could be potentially toxic and should be avoided, especially if taken in combination with diabetes medications, both of which can lower blood sugar.

# Sour and Blue

### By Christian Schaal

This is a variation on the classic "Aviation" cocktail recipe using blueberries. While the Wright brothers developed their flying machine in Kitty Hawk, North Carolina, Elizabeth Coleman White collected wild blueberries from the Pine Barrens in New Jersey to first domesticate the highbush blueberry. Make sure the berries are edible if you are venturing into noncommercially harvested *Vaccinium*—first consult a plant key.

2 oz (60 mL) gin

6 muddled blueberries

1 oz (30 mL) lemon juice

¼ oz (7.5 mL) cherry liqueur

½ oz (15 mL) Simple Syrup
(see page 124)

1 dash Hanging Gardens of
Babylon Bitters (see page 66)
or Mediterranean Garden
Bitters (see page 55)

Extra blueberries
(for garnish)

**Method:** In a shaker, combine all the ingredients and shake with ice. Double strain into a coupe glass. Garnish with blueberries on a cocktail pick. Yields one 4 oz drink.

# Sugar Maple

**Latin Name:** *Acer saccharum* MARSHALL

**Botanical family:**
Aceraceae
(the Maple family)

**Common name(s):**
Sugar maple

**Habitat:**
Native to hardwood forests in Nova Scotia to Quebec, southern Ontario to southeastern Manitoba, and the northern reaches of central and eastern United States.

**Parts used:**
Sap, wood

**Diversity:**
Sugar maple continues to be wild harvested.

**Botany:** Sugar maples often dominate the canopy of their native forests. Flowers are yellow with long pedicels but no petals, with usually only one sex functional within each. The fruit is a double samaras (a flat, papery wing that contains the fruit); the growing season ranges from 80 to 260 days. Seedlings are extremely shade tolerant.

**Cultural Uses:** Native American tribes of the eastern United States used heated stones to boil sap down in a wooden trough, or left it in open vats during a frost, with the ice removed each morning, to concentrate the syrup. The sap is 98 percent water, so it takes about 40 gallons (150 liters) of sap to produce a gallon (3.8 liters) of maple syrup.

**Culinary Uses:** Seedlings are consumed fresh or dried; seed are boiled and eaten hot, the wings removed. The inner bark is dried, ground into powder, and used to thicken soups or with cereals to make bread.

Maple syrup and maple sugar are used widely in cooking.

**Traditional Medicinal Uses:** A tea from the inner bark has been used as a diuretic, tonic, and expectorant. An infusion of the bark has been used to treat blindness, and sap has been used for sore eyes. Maple syrup is used in cough syrups.

**Phytochemistry and Bioactivity:** Fifty-four compounds have been identified (five of which were new to science) that demonstrate anti-inflammatory properties. Antioxidant activity of pure syrup is comparable to strawberry-and-orange juice, and it has a selective antiproliferative activity against certain cancer lines.

**Safety Precautions:** Although maple syrup has lower levels of fructose than many other sweeteners, it is still a sugar and should be consumed in small amounts.

# King of Carrot Flowers

By Kevin Denton

The combination of smoke (aka carbon released into the atmosphere) and maple in this cocktail is ironic. In the northeast United States and Canada, where maples are not only an important crop but an important forest species, the trees have struggled in recent decades, in part because of increasing pollution. With this cocktail, let's raise our glasses to a greener future.

2 oz (60 mL) carrot juice

1 oz (30 mL) tequila

½ oz (15 mL) lime juice

½ oz (15 mL) Smoked Maple Syrup (see page 124)

3 dashes Demon Flower Bitters (see page 64) or Alone in Kyoto Bitters (see page 53)

1 to 2 oz (30–60 mL) tonic soda

**Method:** Add all the ingredients except the tonic to a shaker and then add ice. Shake and double strain into a highball glass with large ice cubes. Top with the tonic. Yields one 5–6 oz drink.

# Osha

**Latin Name:** *Ligusticum porteri* J. M. COULT. & ROSE

**Botanical family:**
Apiaceae (the Parsley family)

**Common name(s):**
Osha, bear root, Porter's licorice root, mountain lovage, chuchupate, loveroot, mountain ginseng, Indian parsley

**Habitat:**
Osha is a perennial herb in the carrot family found in moist, fertile upland meadows throughout its range, encompassing much of the Rocky Mountains from northern Wyoming to Chihuahua, Mexico.

**Parts used:**
Roots and rhizomes, seed, leaves

**Diversity:**
Osha is one of approximately sixty known species of *Ligusticum*.

**Botany:** Osha grows to more than 3 feet 8 inches (100 cm) tall, with lanceolate or elliptic leaf segments and white flowers grouped in compound umbels. The ribbed, oblong red fruit is about ¼ inch (5–8 mm) long. The plant's distinctive taproot has fibrous root hairs and a distinctive odor.

**Cultural Uses:** Some Native Americans burn osha during prayers; the smoke is believed to carry the prayers to the creator. Traditional accounts of bears showing the Navajo osha's healing properties have been substantiated by their use of osha root in captivity and in the wild, captured on stop-motion cameras. Bears chew the root into a mash and spread it on their fur.

**Culinary Uses:** The leaves, seed, and roots can be used to season meat, beans, and chili; the leaves can be boiled and eaten as potherbs or added raw to salads; the roots can be boiled for use in salads and soups.

**Traditional Medicinal Uses:** The roots have been consumed in preparations for colds, coughs, bronchial pneumonia, flu, and other respiratory ailments. Root preparations were used externally to treat pain, digestive problems, scorpion stings, and infections.

**Phytochemistry and Bioactivity:** Forty-two volatile compounds obtained by a solvent extraction of powdered roots have been characterized. The essential oil inhibited a multidrug resistant strain of *Staphylococcus aureus* and contained an antispasmodic.

**Safety Precautions:** There's a risk of product substitution with close relatives, including other *Ligusticum species* and other members of Apiaceae, such as the closely related *Conioselinum scopulorum* or *Conium maculatum* (poison hemlock). Osha may come out in breast milk and potentially increases menstrual flow, so it should be avoided in pregnancy and while nursing.

# This Strange Effect

By Christian Schaal

Osha, also called "bear root," grows wild from Washington to New Mexico, but the plant needs to be harvested by experts, because it looks strikingly similar to poison hemlock. Raw osha root is also potently scented, so it is better to buy proprietary osha honey.

1½ oz (45 mL) Pimm's

½ oz (15 mL) tequila blanco

½ oz (15 mL) lemon juice

¼ oz (7.5 mL) Simple Syrup (see page 124)

¼ oz (7.5 mL) falernum

1 dropper osha honey

3 drops Black Bear's Bitters (see page 65) or Biennial Bitters (see page 54)

4 oz (120 mL) ginger beer

Celery stick, straw, carrot greens, or licorice stick (for garnish)

**Method:** Combine all the ingredients except the ginger beer in a shaker and add ice. Shake and strain over fresh pebble ice into a footed pilsner glass. Top with the ginger beer. Garnish with a celery stick, straw, carrot greens, or licorice stick. Yields one 7 oz drink.

# Lemon Myrtle

**Latin Name:** *Backhousia citriodora* F. MUELL.

**Botanical family:**
Myrtaceae (the Myrtle family)

**Common name(s):**
Lemon myrtle, lemon ironwood, sweet verbena tree

**Habitat:**
Lemon myrtle is a hardy shrub that can be grown in warm temperate to tropical climates but is sensitive to frost during its early growth.

**Parts used:**
Dried, crushed leaves, essential oil

**Diversity:**
Several species in the genus Backhousia have only been described in the last decade and are only known to occur in a single population; all are endemic to Australia.

**Botany:** Lemon myrtle is slow growing but can reach about 15 feet (5 m). The leaves are evergreen, opposite, glossy green, and strongly lemon scented. Small white flowers are produced in clusters; like other species of Myrtaceae, the calyx is persistent after the petals have dropped.

**Cultural Uses:** In the late nineteenth century, citral and citronellal were isolated from the leaves and used as a lemonlike essential oil, and a natural insect repellent. It has been used in the flower industry as a filler in bouquets and is especially prized for its long-lasting leaves and ability to freshen a room.

**Culinary Uses:** Lemon myrtle can be used in flavorings fresh, dried, and as an essential oil. It mirrors the flavor of citrus closely but lacks the associated acidity, making it a particularly useful additive in recipes sensitive to curdling during processing.

**Traditional Medicinal Uses:** The oil is believed to repel fleas and break down the embryonic layer of flea eggs, and it is today used in some chemical-free pet shampoos. It can be used as a repellent against mosquitoes, beetles, cockroaches, houseflies, and dust mites.

**Phytochemistry and Bioactivity:** The oil is mainly citral, and has antiviral properties against herpes, cold sores, warts, and molluscum contagiosum. It can be applied as an antiseptic for stings and certain skin conditions. In commercial use, it is used as a lozenge for sore throats.

**Safety Precautions:** There is no information on adverse reactions of lemon myrtle extracts; concentrated lemon myrtle essential oil has low toxicity for topical use in concentrations above 1 percent.

# The Back House

By Christian Schaal

Lemon myrtle leaf used to be hard to find outside of Australia, but the Internet makes it easy. Many tea suppliers also sell it. If your rum becomes overpowered by the lemon myrtle flavor, you can mix infused rum and plain rum as you like.

1½ oz (45 mL) Lemon Myrtle-Infused Nicaraguan Rum (see right)

1 dash Celery Bitters (see page 51) and 1 dash Lazy Hippie One-Day Bitters (see page 52)

½ oz (15 mL) falernum

¾ oz (20 mL) lime juice

½ oz (15 mL) Italian herbal liqueur

¼ oz (7.5 mL) Simple Syrup (see page 124)

Lime zest (for garnish)

**Lemon Myrtle-Infused Nicaraguan Rum:** Combine ¾ tsp lemon myrtle leaf with 6 oz (175 mL) Nicaraguan rum in a blender and blend for 30 seconds. Let sit for 15 minutes, then strain. Yields about ¾ cup (175 mL).

**Method:** In a shaker, combine all the ingredients and then add ice. Shake and fine strain into chilled coupe glass. Garnish with lime zest. Yields one 4 oz drink.

# Gum Trees

**Latin Name:** *Eucalyptus* spp.

**Botanical family:**
Myrtaceae (the Myrtle family)

**Common name(s):**
Eucalyptus, red gum, blue gum, lemon-scented gum, gum trees, olida, ironwoods, mallee

**Habitat:**
Semiarid and woodland climates in Australia, with some in arid regions.

**Parts used:**
Leaves, stem wood, bark

**Diversity:**
There are approximately eight hundred species, all native to Australia.

**Botany:** Eucalyptus are fast-growing large trees or shrubs that can live for five hundred years. The leaves are often waxy blue-gray or glossy green. The flowers have showy stamens and a single central style. The bark is diverse in color, layers, and texture. Most gum trees are fire tolerant, and may even promote fire by creating dense mats of oil-rich leaves and capsules.

**Cultural Uses:** Trees are used mainly in timber and paper production, but they are important resources for nectar, as a shelter for animals, as a fuel, and an essential oil. All parts are used in essential oil extraction.

**Culinary Uses:** Australian indigenous people consume the fruit of a few species. Flower nectar can be sucked or made into drinks. Galls made by insect infestations are a popular food, as are the grubs and bees' nests hosted by many trees. Sap or gum is also edible from some species.

**Traditional Medicinal Uses:** Leaves, stems, and bark are extracted and made into an antiseptic to wash sores and cuts, and they are also used as a warm body wash to relieve muscle aches, pains, and inflammation. Water extract of the inner bark has been used to treat toothaches and clean teeth.

**Phytochemistry and Bioactivity:** Contains numerous volatile compounds with antiseptic properties; the dominant volatile is ether 1,8-cineole, known as eucalyptol. These volatiles have been shown to stimulate mucous-secreting cells in the nose, throat, and lungs.

**Safety Precautions:** Toxicity can result from ingesting or topically applying higher than recommended doses of eucalyptus oil. Dogs and children should avoid eating the leaves or handling them, because they can cause skin irritation.

# The Roving Duck

By Christian Schaal

Eucalyptus, also known as gum trees, are everywhere. Most species are used as a healing tea and can be used in cocktails, especially the species called red gum and silver dollar gum. The flavor will change when extracting in alcohol, and they should be consumed in small quantities.

1½ oz (45 mL) Eucalyptus-Infused Vodka (see right)

1 oz (30 mL) Persian Cucumber Juice (see right)

½ oz (15 mL) lime juice

½ oz (15 mL) Simple Syrup (see page 124)

5 drops Mount Apo Bitters (see page 68) or Bienniel Bitters (see page 54)

Eucalyptus leaf (for garnish)

**Eucalyptus-Infused Vodka:** First estimate the amount of vodka needed for the number of cocktails you plan to make. Blend 6 dried small eucalyptus leaves for every 6 oz (175 mL) vodka until homogenized (uniformly mixed and incorporated). Let sit for 5 minutes, then strain. Yields about ¾ cup (175 mL).

**Persian Cucumber Juice:** Avoid cucumbers with a wax coating: in this drink, we prefer the taste of unpeeled cucumbers but don't like the wax. Slice a cucumber lengthwise to remove the seeds. Blend the cucumber in as much water as needed to completely homogenize. For us, it was 6 oz (175 mL) water for one medium cucumber. Use immediately.

**Method:** In a shaker, combine all the ingredients, add ice, shake, and double strain into a Nick and Nora glass. Garnish with a eucalyptus leaf. Yields one 3½ oz drink.

# Yams

**Latin Name:** *Dioscorea* spp.

**Botanical family:**
Dioscoreaceae
(the Yam family)

**Common name(s):**
Yams, greater yam, purple yam, winged yam,
white yam, Guyana arrowroot

**Habitat:**
Tropical regions, particularly in Africa and the Pacific Rim.

**Parts used:**
Tuber, young shoot

**Diversity:**
There are more than six hundred species in the
*Dioscorea* genus; only a few species are edible.

**Botany:** All yam species are perennial herbaceous plants. They have tuberous roots and heart-shaped leaves. Some tubers grow underground while others develop above ground. Tubers can weigh up to 22 pounds (10 kg). The fruit is a brightly colored capsule, usually about 1¼ inches (3 cm) in diameter.

**Cultural Uses:** Some regions of the world rely heavily on yams as a main source of calories; most cultivated species, when cooked, are delicious. There is incredible diversity in the culinary preparations of the tubers and shoots; in religious uses of the plants as offerings and part of religious rites; and in traditional and cutting-edge therapies.

**Culinary Uses:** Tubers have a high caloric value due to their starch content. Purple yam tuber (*D. alata*) is used in desserts and to flavor ice cream and milk products. Many species have bitter-tasting shoots used to enhance soups.

**Traditional Medicinal Yses:** Yams have been widely used to regulate estrogen levels and are used to treat cramps, coughs, hiccups, muscle spasms, and gas. Root decoctions have been used to treat rheumatism and arthritis, bowel irritation, labor pains, and nausea. Tuber paste is used topically to treat flaky skin.

**Phytochemistry and Bioactivity:** Tuber consumption can lower cholesterol. Yams are high in the B vitamins, which helps battle fatigue, depression, and anxiety. They are high in fiber, low in fat, and low protein.

**Safety Precautions:** Yams contain steroidal saponins that can be toxic in high doses. Many species have tubers that must be boiled or soaked to remove the toxins. Some species have high amounts of calcium oxalate, crystals that can be an irritant. Yam leaves also contain these crystals, so should be cooked before consumption, too.

# Toto Te Taro

### By the Shoots and Roots Team

Eastern Polynesians brought many plants with them to New Zealand when they first arrived about eight hundred years ago, and the Maori culture that subsequently developed continues to make heavy use of these tuber crops. So we name the drink after taro and ask it to be made from ube and yam. You can interchange them, depending on what you have available, or substitute sweet potato for a slightly sweeter cocktail.

---

1½ oz (45 mL) Ube- and Yam-Infused Rum (see right)

4 oz (120 mL) Basque cider

½ oz (15 mL) triple sec liqueur

1 dash Fruitbelt Bitters (see page 57) or Thai Market Bitters (see page 53)

Sea salt

Yam leaf, orange peel (for garnish)

**Ube- and Yam-Infused Rum:** Oven roast or barbecue whole in aluminum foil about 9 oz (250 g) each of ube and yam for up to about 45 minutes depending on your cooking method, until cooked through. When done, you should be able to easily pierce them with a fork, like you do with a baked potato. Let cool, remove the peel, and blend with white rum until they make a slurry with the alcohol. Let sit overnight to settle. Pour off and retain the top layer, which will look conspicuously different from the starch layer that has settled on the bottom, and fine strain through a coffee filter or a nut milk bag. Use a centrifuge if you have one.

**Method:** Stir the rum, cider, and triple sec, pour into a Collins glass over ice, and add the bitters. Garnish with a yam leaf, if available, an orange peel twist, and a sprinkle of salt. Yields one 6 oz drink.

123

# Syrups

Although we try to dial down the sugariness of drinks, we still love all the different sweeteners that can form the basis of syrups. Syrups are important in cocktails and complement the bitters. You can easily infuse glycerine, maple syrup, birch syrup, diluted honey, coconut, date syrup, sorghum syrup, or simple syrups with whatever botanical you choose.

Each different sugar chain has a different effect on how a beverage is perceived. Most recipes in this section use basic granulated sugar derived from sugar beet or sugarcane, but you could easily modify the recipe to infuse directly into some of the more exotic options. Just be aware that the color of some of these will not always make your drinks attractive.

Sugars are bacteriostatic, meaning they inhibit bacteria from growing and spoiling food. The more concentrated the syrup is with sugar molecules, the longer it will last. To err on the side of caution, we recommend all syrups are stored in the refrigerator and used within a couple of weeks.

### Simple Syrup

This is a staple ingredient of many cocktails.

½ cup (100 g) granulated sugar

½ cup (120 mL) water

**Method:** Combine the sugar and water in a saucepan, then heat over medium, stirring, for a few minutes until all the sugar has dissolved. Remove from the heat and let cool. Yields 150 mL.

### Rich Syrup

Make a Simple Syrup but use 1 cup (200 g) of sugar instead of ½ cup (100 g). Yields 150 mL.

### Mango Black Tea Syrup

Make a Simple Syrup but add 1 tablespoon mango black tea to the water as you heat the syrup. Strain the syrup before bottling.

### Honey Syrup

Make a Simple Syrup but use ½ cup (170 g) of honey instead of ½ cup (100 g) of sugar.

### Smoked Maple Syrup

You can create your own Smoked Maple Syrup using a barbeque or on the stovetop. It is used in the King of Carrot Flowers cocktail (see page 115).

2 cups (630 g) maple syrup

2 handfuls oak chips, to about ½ inch (1 cm) in depth

**Method:** If using a barbeque, heat the coals to 225°F (107°C). Soak the oak chips in water, then put them into a smoking basket in the barbeque (follow your brand's directions). Pour the maple syrup into a heatproof dish or skillet and place the dish on the barbeque. Cover and let smoke for 20 minutes. Pour the Smoked Maple Syrup into a jar and store in a cool place.

Alternatively, warm the maple syrup in a saucepan over low heat and blend in one part smoky mezcal to four parts maple. Stir slowly to boil off some of the alcohol but be careful not to let the maple burn. It should be hot to the touch but not bubbling. Remove from the heat and let cool. Yields 175 mL.

## Oolong Tea Simple Syrup
This is an essential ingredient for the Green Glory cocktail (see page 75).

2 tsp oolong tea leaves

1 cup (240 mL) water

1 cup (200 g) granulated sugar

**Method:** Bring the water and leaves to a boil in a saucepan over low heat, then simmer for 10 minutes. Keep adding water if the level drops. Add the sugar and stir until dissolved. Remove from the heat and let cool, then strain through a nut milk bag or French press and discard the solids. Yields 350 mL.

## Chinese Five-Spice Syrup
A deliciously complex flavor used in the Wu Wei Mei cocktail (see page 79).

1 Tbsp Chinese five-spice powder

1 cup (240 mL) water

1 cup (200 g) granulated sugar

**Method:** Lightly toast the spice powder in a saucepan, then add the water and sugar. Bring to a low boil and stir until the sugar is completely dissolved. Remove from the heat, let cool for 10 minutes, and strain through

a French press or nut milk bag. Let stand for 15 minutes and strain again. Yields 350 mL.

## Schisandra Syrup
A favorite of ours, because it's part of our cocktail tribute to our hero, Nicolai Vavilov (see page 81).

4 Tbsp dried schisandra berries

1 cup (240 mL) water

½ cup (110 g) honey

**Method:** Mix the berries with the water in a saucepan. Bring to a boil over low heat, then simmer until reduced by half, remove from heat, and add the honey. Let cool to room temperature. Pour the mixture into a blender and blend on high for 30 seconds. Let stand for 15 minutes, then strain through a nut milk bag or French press and discard the solids. Yields 350 mL.

## Baobab Syrup
A key ingredient in the Bab's Dock cocktail (see page 85), but one you can mix in with your own recipe inventions. It works wonders in blended ice drinks.

1 tsp baobab fruit powder

⅜ cup (90 mL) Simple Syrup (see page 124)

**Method:** Combine all the ingredients in a shaker and shake vigorously. Use immediately. Yields 90 mL.

Schisandra

125

## Lucuma Simple Syrup

An easy recipe with a yummy raisin flavor.

1 tsp lucuma fruit powder

¼ cup (60 mL) Simple Syrup (see page 124)

**Method:** Combine all the ingredients in a shaker and shake vigorously. Use immediately. Yields 60 mL.

## Fire-Roasted Carob Syrup

Carob (*Ceratonia siliqua*) is a tree from the Mediterranean region with long bean pods that have chocolaty notes. We suggest mixing this syrup with Amontillado sherry or cachaça or using it in recipes in place of crème de cacao.

1 carob pod

⅞ cup (200 mL) water

1⅛ cups (230 g) granulated sugar

**Method:** Crack the carob pod before roasting over a flame. We put it directly over a gas range and rotate it like a kabob every minute until all sides have been darkened and a chocolaty aroma permeates the kitchen. In a powerful blender, blend the carob pod into the water. Transfer the mixture to a saucepan, add the water, and bring to a boil over low heat, then add the sugar and stir until it has dissolved. Simmer for 20 minutes. Remove from the heat and let stand for an hour or longer until the desired strength—as it sits, it will become more bitter. Double strain through a fine mesh strainer and a French press or nut milk bag. Yields 280 mL.

Vanilla

## Bitter Walnut Syrup

This is a quick and simple method to make Nocino, a surprisingly spicy and flavorful liquor originating in Sicily. We think it works as a great bitters for a Manhattan.

15 to 20 young green unripe walnut fruit

10 whole cloves buds

½ vanilla bean

Zest from 1 lemon

3 Tbsp lemon juice

2⅛ cups (500 mL) water

2½ cups (500 g) granulated sugar

**Method:** Wearing gloves, cut the walnut fruit in half. Soak them in room temperature water, changing the water daily, for 5 days to partly remove the bitterness or for 10 days to completely remove the bitterness.

Combine the walnut fruit, sugar, and water in a saucepan, then bring to a boil over low heat and simmer for 30 minutes. Add the remaining ingredients and simmer for another 30 minutes. Let cool, then strain through a nut milk bag or French press and discard the solids. Transfer to a jar and store in the refrigerator for a week before use. Yields 750 mL.

## Saffron Syrup

A particularly precious spice, used in the Superbloom cocktail (see page 93).

10 stigmas saffron

½ cup (100 g) sugar

½ cup (120 mL) water

**Method:** Combine all the ingredients in small saucepan, then bring to a boil over a low heat and simmer, stirring, for 5 minutes. Cool, then strain through a nut milk bag or French press and discard the solids. Yields 150 mL.

Honeybush

## East West South: African Tisane Syrup

Africa's astounding botanical diversity has been used in tisanes and soups to aid healing for thousands of years. We used this syrup with Colorado barley whiskey, replacing the sugar and bitters when making an old-fashioned.

120 mL water

10 g dried red rooibos leaves

10 g dried honeybush leaves

10 g dried coffee leaves

5 g round-leaf buchu*

1 broken prekese fruit pod

100 g sugar

**Method:** Bring the water to a boil in a saucepan over low heat and add the rooibos, honeybush, and coffee leaves. Simmer for 15 minutes, then add the buchu, prekese, and sugar. Let cool, then strain through a nut milk bag or French press and discard the solids. Yields 150 mL.

*This species is still wild harvested, but harvest is currently outpacing reproduction of the plant. Please check your source to make sure they are making improvements toward sustainable harvest and assistance.

## Curry Leaf Apple and Kampot Pepper Syrup

The Kampot region of Cambodia has been growing this pepper with knowledge passed down since the thirteenth century. Curry is used in countless treatments to help heal and strengthen many parts of the body. This syrup is great as a base for sodas, salad dressing, and cooking, or it can be taken straight if you need a pick-me-up. Special thanks to Dr. Michael Purugganan, who introduced the Kampot pepper to Shoots & Roots.

1 tsp Cambodian kampot black peppercorns

6 fresh curry leaves

1 tsp malic acid powder

½ small Granny Smith apple

½ cup + 1 Tbsp (115 g) sugar

⅓ cup + 1½ Tbsp (100 mL) water

**Method:** Blend all ingredients for 1 minute and let rest for 30 minutes. Strain and bottle. Unlike the other syrups, this syrup should keep for around a month in the refrigerator. Yields 150 mL.

Pepper

# Using Bitters in Other Dishes and Drinks

Bitters can be used in countless ways, from everyday beverages to unique cocktails and as flavor additions in food. You can think of them as a liquid spice blend; we have applied them with great versatility into diverse dishes and drinks. The following are basic recipes with bitters. The drink recipes in this section make one drink that is 2 to 8 ounces (60–240 mL) in size.

For these basic drink recipes, the bitters should be mixed in with the other ingredients in one of four ways. You can add bitters to your glass first, letting the other ingredients mix everything together. The bitters can also be added and then swirled around to coat the inside of your glass. You can add the bitters after all the other ingredients have been poured, and then stir your drink. Finally, the bitters can be added in a cocktail shaker with the other ingredients and shaken; the use of cocktail ice is optional. You can top off your drink with a garnish of your choice (see box opposite).

## Bitters and Soda

2 to 3 dropperfuls of bitters

½ cup (120 mL) chilled sparkling water or club soda

Optional: ¼ cup (60 mL) fresh juice or coconut water

Optional: Sweetener to taste (maple syrup, honey, sugar syrup, agave, stevia, or monk fruit)

## Bitters and Tea and Tisanes

2 to 3 dropperfuls of bitters

1 cup (240 mL) hot, iced, or cold-brew tea or tisane

Optional: Sweetener to taste (maple syrup, honey, sugar syrup, agave, stevia, or monk fruit)

## Bitters and Coffee

2 to 3 dropperfuls of bitters

1 cup (240 mL) hot, iced, or cold-brew coffee

Optional: Sweetener to taste (maple syrup, honey, sugar syrup, agave, stevia, or monk fruit)

## Bitters and Hot Chocolate

2 to 3 dropperfuls of bitters

1 cup (240 mL) hot chocolate

Optional: Sweetener to taste (maple syrup, honey, sugar syrup, agave, stevia, or monk fruit)

## Bitters and Coconut Water

2 to 3 dropperfuls of bitters

1 cup (240 mL) chilled coconut water

## Bitters and Juice

2 to 3 dropperfuls of bitters

1 cup (240 mL) freshly squeezed juice

## Bitters and Gin

2 to 3 dropperfuls of bitters

¼ cup (60 mL) gin

Optional: Sweetener to taste (maple syrup, honey, sugar syrup, agave, stevia, or monk fruit)

Optional: ¼ cup (60 mL) fresh juice

## Bitters and Whiskey

2 to 3 dropperfuls of bitters

¼ cup (60 mL) whiskey

Optional: Sweetener to taste (maple syrup, honey, sugar syrup, agave, stevia, or monk fruit)

## Bitters Spritzer

2 to 3 dropperfuls of bitters

¼ cup (60 mL) chilled sparkling water

¼ cup (60 mL) white or rose wine

## Bitters Salad Dressing

6 dropperfuls of bitters

2 Tbsp shallots, finely chopped

2 Tbsp red or white wine vinegar

2 tsp mustard

6 Tbsp extra virgin olive oil

Pinch of sea salt

Pinch of black pepper

## Bitters in Baking

Replace vanilla and other extracts with bitters in baking recipes, such as cakes and cookies.

## Bitters and Ice Cream

1 scoop of ice cream, sherbert, yogurt, or sorbet

1 to 2 dropperfuls of bitters drizzled on top

## Bitters Marinade

2 dropperfuls of bitters per 1 oz (25 g) of raw meat, fish, or tofu

Marinate for at least 4 hours before cooking.

---

## GARNISHES TO TOP OFF DRINKS

01. Citrus peels, slices, zest, twists, and wedges

02. Herb sprigs, such as lavender or thyme

03. Edible leaves, such as mint or basil

04. Berries

05. Fruit slices, such as apple or pineapple

06. Vegetable slices, such as cucumber or celery

07. Edible flowers

08. Seeds, such as pomegranate seeds or basil seeds

09. Fresh bark sticks

10. Edible rims, such as sea salt or chili powder

# Shrubs

**Recipes by Jim Merson**

Shrubs are a way of preserving fruit from rot, and ideally the fruit should be ripe but used right before it becomes too ripe. To make one of these shrub recipes, which will take about 30 minutes, the fruit is mixed with vinegar, strained, and then sweetened. This mixture is then used to make a refreshing beverage.

A shrub complemented by a fresh garnish is a way you can enjoy bitters on a daily basis. Shrubs are also an excellent way to learn about botany and microbiology. To make them is to tune your attention and your senses.

Play it safe. Shrubs rely on the vinegar to preserve the fruit—the high acidity helps keep pathogens away. However, because the process described here for making shrubs will not heat the mixture up enough to kill any probiotics (that is, beneficial bacteria), they may not kill harmful bacteria either, and the recipes have a hefty dose of fruit and are not strained much. All of this makes spoilage more probable, so once you make a shrub, store it immediately in the refrigerator and use within a month or two.

These recipes all use the same kind of vinegar and sweetener. I favor agave syrup because it tends to taste sweeter per volume than simple syrup or honey, and it combines easily. Some people like it because of its low glycemic index, too; it's a healthy as well as a tasty option.

## Method

Prepare the fruit as necessary, washing, stemming, seeding, or pitting it. Then puree it with apple cider vinegar in a powerful blender for a minute or two, making sure that the pitcher doesn't get too warm or lose its lid.

When blended, pour the puree through a metal chinois strainer into a thick plastic 4-qt (4.5-L) graduated pitcher. You can help the liquid pass through by tapping the chinois with your palm or by stirring and forcing the puree gently through with a large ladle. Add the agave syrup to the filtered puree and stir to combine.

Taste the mixture; it will be pretty intense the first couple of times you do it. The bite from the vinegar when the mixture is correct should be right at the point of being gentle, still present but not too intense. If there's too much bite, add a little more agave. Tweak with small quantities, just an ounce or two at a time. It should be sweet but still taste like fruit and vinegar. You'll be serving the shrub with sparkling water, which contains carbonic acid that will create a more acidic taste overall, so don't let the sweetness throw you off.

When you think you've found a balance of a nice mellow bite, measure out 3 tablespoons (45 mL) of syrup into a glass with ice and add ⅞ cup (200 mL) of sparkling water to it. Stir vigorously and taste. Does it need tweaking? Give it to either your least favorite or most trusted friend in the room to get an opinion. If both are in the room, just give it to one and keep them both guessing where they stand with you. Friends love a mystery.

## Monterey Strawberry Shrub

The key to making a great strawberry shrub is to find a way to communicate the beautiful fragrance of the strawberries through the din of the vinegar. Let your strawberries get right on the verge of being overripe before you use them; the ideal strawberry for a shrub should be wrinkly. It may have an off-putting texture, but we're about to eliminate the strawberry's texture by pureeing it.

This was the first shrub I learned to make for that reason, and I think it's a great introduction to shrub making. You get to practice dealing with more of a fragrance than a flavor. With a bunch of really ripe strawberries, this recipe shouldn't need too much dialing in at all. White strawberries are not recommended for shrubs.

2 cups (300 g) Monterey strawberries (or other red variety), stemmed

1¼ cups (300 mL) unfiltered apple cider vinegar

1 cup + 3 Tbsp (260 g) agave syrup

**Method:** Make the shrub according to the method above.

**To serve:** Pour 1½ oz (45 mL) Monterey Strawberry Shrub over a glass with ice and top with about ⅞ cup (200 mL) of cold sparkling water. Stir vigorously and garnish with a fresh strawberry.

## Peach Shrub

Finding balance in a peach shrub requires more of a delicate hand to expose the richness and perfume. Peaches are so flavorful, especially the really ripe ones that are used for making shrubs, but their richness can be buried by the vinegar funk. Just the right amount of vinegar will make your peaches sing with summery joy.

Yellow or white peaches come in many varieties that arrive at different times of the season. I envisioned yellow peaches as I was writing out this recipe, but this recipe will work for making a marshmallowy sweet white peach shrub as well. Naming your shrub after the variety of peach you use makes things even more fun. Names like July Flame Shrub or Arctic Queen Shrub are a slam dunk.

2 cups (310 g) pitted and sliced peaches

Scant 1¼ cups (290 mL) unfiltered apple cider vinegar

1 cup + 3 Tbsp (260 g) agave syrup

**Method:** Make the shrub according to the method above.

**To serve:** Pour 1½ oz (45 mL) Peach Shrub over a glass with ice and top with about ⅞ cup (200 mL) of cold sparkling water. Stir vigorously and garnish with a sprig of mint.

Peach

## Bartlett Pear Shrub

Superripe Bartlett pears mingle with apple cider vinegar and agave like no other fruit to really earn the descriptor *juicy*. Pears are unique in the fruit world for having a particular cell structure called *sclereids* that won't break down in a puree. The result is a soda that will line your glass in little specks of pear. You can feel these little specks on your lips when taking a bite into a ripe pear. If you run your tongue along the roof of your mouth immediately after swallowing that bite, you can feel them, too. You will hardly notice them in the drink until it comes time to do the dishes.

2 cups (280 g) stemmed, cored, and sliced Bartlett Pears

1¼ cups (280 mL) unfiltered apple cider vinegar

1 cup + 3 Tbsp (260 g) agave syrup

**Method:** Make the shrub according to the method on pages 130–31.

**To serve:** Pour 1½ oz (45 mL) Bartlett Pear Shrub over a glass with ice and top with about ⅞ cup (200 mL) of cold sparkling water. Stir vigorously and garnish with a sprig of mint.

Apple

## Kumquat Shrub

This sweet and tangy shrub has a serious citrus bite and an incredible lingering aftertaste. Kumquats make for a funky shrub, zippy and slightly bitter from the citrus oil. Kumquats are best used when fresh. Their tanginess can mellow as they ripen, but they have the tendency to simply dry out as they age. You want to use the acidity in the fruit to substitute the acidity in the vinegar for the flavor profile of this shrub.

Stems from the kumquats come off easily with your fingers. You can de-seed kumquats by slicing them into kind of a spiral with a paring knife and poking the seeds out; they have six or so seeds on average. Then the whole fruits are popped into the blender, peel and all. The skin of the kumquat has both the wonderful citrus oil that will make for this shrub's incredible aftertaste but also all of the sweetness in the fruit. Puree and strain, but reserve the solids to introduce back into the syrup depending on how much sweeter it needs to get. Keep some peel/puree in the final strained drink for texture and sweetness, but don't be shy to let agave do the leg.

1¼ cups (170 g) stemmed and seeded kumquats

1 cup (240 mL) unfiltered apple cider vinegar

1¼ cups (280 g) agave syrup

**Method:** Make the shrub according to the method on pages 130–31.

**To serve:** Pour 1½ oz (45 mL) Kumquat Shrub over a glass with ice and top with about ⅞ cup (200 mL) of cold sparkling water. Stir vigorously and serve without a garnish.

## Blackberry Shrub with Lemon Balm

Bright sweet blackberries are complemented by a tangy, grassy tickle on the nose from lemon balm. Blackberries have the tendency to go from very ripe to rotten really quickly. Finding that sweet spot can be difficult, so I recommend using them as soon as you get them. The same goes with any recipe with fresh herbs or vegetables. Using fresh herbs right away means you'll be using them while they still have all of their potency. The flip side is that using blackberries so quickly will mean that they have more acidity and possibly underdeveloped sweetness compared to the way a superfragile overripe blackberry tastes. To compensate, this recipe calls for slightly less vinegar than is used in the strawberry shrub.

Puree lemon balm leaves with the berries and apple cider vinegar until fine enough for some to pass through the mesh of the chinois strainer. You'll want to strain out the blackberry seeds, but in doing so, you will strain out some of the leafy lemon balm, however, a lot of the leaves and their essential oils will pass through. The bright acidity of the fresh berries carries over the vinegar while the tangy grassy lemon balm greets the nose and again on the aftertaste.

2 cups (290 g) blackberries

1¼ cups (290 mL) unfiltered apple cider vinegar

1 cup + 3 Tbsp (260 g) agave syrup

15 g fresh lemon balm leaves, plus extra for garnish

**Method:** Make the shrub according to the method on pages 130–31.

**To serve:** Pour 1½ oz (45 mL) Blackberry and Lemon Balm Shrub over a glass with ice and top with about ⅞ cup (200 mL) of cold sparkling water. Stir vigorously and garnish with three or four leaves of lemon balm. Give them a good smack by clapping them between your hands to bring out the essential oils in the leaves. The aroma of the freshly clapped leaves and the lemon balm in the syrup should be potent enough to add to the bouquet of berries and vinegar.

Lemon balm

Chapter 6
# The Plant Directory

We hope this book helps increase your interest in exploring plants and creating your own bitters concoctions. Here, we've organized some of our favorite plants for bitters and beverages by geographic region and given you a cheat sheet of some interesting information about them. We encourage you to do like botanists and explore the plants presented here by creating your own bitters concoctions and cocktails. You may want to mix botanicals from a certain region, form, use, or flavor. Our plant directory is preceded by directions on making your own bitters. Happy explorations!

# Keeping Records

Getting into the habit of taking detailed notes at all stages of your bitters-making process can provide valuable information for later on. These records will be especially useful if you want to repeat a formula or make changes to it to suit your own tastes. Make sure that you note not only what did work but also what didn't—to avoid repeating past mistakes.

The notes you take should help you to remember the attributes of plants that you may forget later on, such as the fresh appearance or aroma of the plant. Make sure you carefully record all important details, such as the accurate measurements of the quantities used, the dates of different steps, and the information of suppliers or location of collection. You could make your own chart to help organize your notes. Most critically of all, don't forget to label your jars and bottles with the contents and date them.

## Botanical Nomenclature

**Dandelion** *Taraxacum officinale* (L.) WEBER EX F. H. WIGG.

| Common Name | Genus | Species | Original Authority | Current Authority |
|---|---|---|---|---|

The plants in the directory follow the International Plant Names Index naming convention, listing the Latin genus name and species. Note that common names can differ not only in different countries but in different regions in the same country.

If you find a plant that has been reclassified, the original author's name may appear in parentheses. "Ex" is used if one author has expanded upon the original author's description, while "and" indicates joint authorship.

# Transcontinental

### Gentians *Gentiana* spp.

Gentians are iconic symbols for myths and values of the mountains, and they are even used in the Alps to prevent loss of appetite and protect the liver and gall from disease. Gentian can alter blood pressure. A few toxic plants, such as hellebore, resemble gentians, so make sure you know what you're picking if you are doing the harvesting.

**Plant part used:** Roots

### Mints *Mentha* spp.

Most species in the genus *Mentha* occur wild and are cultivated in the United States, Europe, and Africa. The highly scented leaves contain up to 5 percent essential oil, which is used to treat many conditions including digestive problems, nausea, and fever. Season, stress, and plant age affect the aromatics—first try the mint and build from there.

**Plant part used:** Leaves

### Citrus/Orange *Citrus* spp.

All cultivated citrus are either natural or artificial hybrids that have occurred over the last several thousand years between four true citrus species: citron, mandarin, pomelo, and papeda. Today's citrus fruits are related in a complex pedigree of crosses and are a source of many beneficial bioactive compounds, including amino acids, essential oils, flavonoids, and vitamin C.

**Plant part used:** Fruit

### Agave *Agave* spp.

Standing solemn on the arid landscape of Mexico and the American Southwest, the agave is a fierce plant. While technically perennial, they only live until sending out their flowering stalks to cast seeds across the desert floor. The buildup of sugar and carbohydrate-rich sap required for this flowering is also the source of agave syrup, tequila, and mescal.

**Plant part used:** Sap, flowers, and leaves

## Cherry *Prunus* spp.

*Prunus* is a genus of about four hundred stone fruits originating in Eurasia, of which cherry is not one species but many. *Malhab*, roasted pits of the mahaleb or St. Lucie cherry (*P. mahaleb*), is a favorite spice of ours for making carbonated beverages. From Turkey to Sudan, all parts are used as medicinal ingredients for the heart and stomach.

**Plant part used:** Fruit

## Rhubarb *Rheum rhabarbarum* L. and *Rheum palmatum* L.

Rhubarb is prized for its fleshy stalks with tart, tangy flavor and avoided for its toxic leaves that are rich in oxalic acid, a nephrotoxic (damaging to the kidneys) and corrosive acid. The roots are used in traditional Chinese medicine as a laxative. Rhubarb has also been used in Arabic and European medicine.

**Plant part used:** Stalks

## Celery *Apium graveolens* L.

An extremely versatile plant with all parts being used as food and medicine, the use of celery dates back to antiquity. The leaf stalk (petiole) is what we commonly consider when we think of celery, however, the underground portion (hypocotyl) is consumed and the leaves and seeds are used as a seasoning. Celery oil is used in perfumery for a floral note.

**Plant part used:** Stalks and leaves

## Coriander *Coriandrum sativum* L.

A plant of the Old Testament and associated with tombs, coriander has been the base of many ancient food, beverage, and ceremonial recipes. Plant-breeding programs have great potential to improve the quality of the herb, known as cilantro, and diversify flavors—but little can be done for the about 14 percent of people who dislike its taste.

**Plant part used:** Leaves and seed

## Cumin *Cuminum cyminum* L.

Cumin seed have long been valued as a digestive aid in South Asia and the Middle East, so much so that its Sanskrit name translates to "that which helps digestion." Cumin seeds are nutritious and among other health attributes, support immune function. Cumin has moderate interactions with several medications so should be taken with caution.

**Plant part used:** Seed

### Licorice *Glycyrrhiza glabra* L.

Licorice has been used widely as a natural toothpaste by chewing on the roots. Not only is it an effective antiseptic and anti-inflammatory, it sweetens the breath, too. The word "licorice" is used to describe the flavor of many diverse plants containing anethole, a widely occurring compound also found in anise, fennel, and camphor.

**Plant part used:** Roots

### Anise *Pimpinella anisum* L.

Anise, or aniseed when referring to the seed, was traditionally a Mediterranean crop that has largely had its center of production moved to the Middle East. Anise is widely used in baking and in beverage flavoring around Europe and the Middle East, with liquors made from anise including absinthe, aquavit, sambuca, and more. Anise is considered a diuretic and diaphoretic (induces perspiration).

**Plant part used:** Seed

### Savory *Satureja montana* L.

Leaves of summer savory can be used as a flavoring or garnish for cooked foods and impart a tangy, peppery flavor reminiscent of marjoram. The plant repels insects and is useful as a biological control in herb gardens. Essential oils show a broad spectrum of antimicrobial activity against yeast species.

**Plant part used:** Leaves

### Hibiscus *Hibiscus* spp.

Hibiscus refers to several hundred species of ornamental flowers native to warm temperate, subtropical, and tropical regions worldwide. It is consumed for its tangy flavor, which pairs well with many ingredients. Dried hibiscus flowers are used in various countries around the world to make a vibrant-colored tisane.

**Plant part used:** Flowers

### Lemongrass *Cymbopogon* spp.

*Cymbopogon citratus*, Western lemongrass, is the most widely cultivated, naturally occurring from India to Malaysia and used in food and medicine in South and Southeast Asia. Other species can be frost and fire resistant.

**Plant part used:** Stems and leaves

### Rose *Rosa* spp.

There are several hundred species of the genus *Rosa* that have petals, buds, and hips (fruit) used in food, beverages, and medicine. Rose petals are used to make tisanes, tonics, rosewater infusions, and syrups that are extensively used in Middle Eastern, Persian, South Asian, and French cuisine. Vodka brings out a metallic quality in rose.

**Plant part used:** Fruit and flower petals

### German Chamomile *Matricaria chamomilla* L.

The root of the word "chamomile" comes from the Greek *kamai melon*, for ground apple. Chamomile is taken as a tea to treat insomnia, relieve anxiety, and stimulate digestion. A mild, sweet taste and slight bitterness, paired with its many beneficial properties, make German chamomile an excellent addition to almost any bitters blend.

**Plant part used:** Flowers

### Lemon Balm *Melissa officinalis* L.

Lemon balm was being used as early as six thousand years ago around what is now Zurich, but the story goes back much farther. Hints of its peculiarities are evident in ancient Turkey, where it was grown next to hives to keep bees returning. It is suggested that before *homonins* had fire, they rubbed lemon balm onto their skin to prevent bees from becoming aggressive.

**Plant part used:** Leaves

### Ginseng *Panax* spp.

Ginseng is native to parts of North America and eastern Asia, where it is used in traditional medicine. The roots are widely infused as tonics and tisanes. Chronic use can cause unwanted side effects; there is also a low risk of adverse interactions with certain medicines, including warfarin, phenelzine, imatinib, and lamotrigine.

**Plant part used:** Roots

### Thuja *Thuja* spp.

An essential oil made from the young leaf tips of thuja is administered to support antibiotic treatments for severe bacterial infections, viral infections, and as a general antiseptic. The compound thujone, present in the essential oil, is toxic to humans in large doses, but it can be an effective remedy for warts and polyps when applied topically.

**Plant part used:** Leaves

### Lovage *Levisticum officinale* W. D. J. KOCH

Since ancient times, lovage roots have been used in Europe to increase alertness and for topical sores, kidney stones, malaria, and more—it was even used as a component of love potions. In culinary applications, lovage is like celery on steroids; it is so complex that any herb paired with it pulls out an interesting complementary note.

**Plant part used: Leaves**

### Dandelion *Taraxacum officinale* (L.) WEBER EX F. H. WIGG.

Dandelion is a perennial herb native to the western hemisphere, with a widespread history of traditional use in treating liver and kidney diseases and in enhancing the immune response against upper respiratory tract infections. The root is bitter; dried root is used in many detoxifying blends. Dried leaves are bitter with a grassy note.

**Plant part used: Leaves, flower buds, flower petals**

### Sloe *Prunus spinosa* L.

When making sloe gin, berries are typically harvested after a frost, then punctured (traditionally with a spine from a blackthorn bush) and coated in sugar before being put into a container with gin, agitated for several months, and strained. Sloe berries are used to treat inflammation of the mouth and pharynx, and the pulped fruit can be applied as a face mask.

**Plant part used: Fruit**

### Angelica *Angelica archangelica* L.

The sweetly scented roots and stems are a flavoring agent and have medicinal properties. Found wild in Russia, Finland, Sweden, Norway, Denmark, Greenland, the Faroe Islands, and Iceland, the roots, fruit, flowers, and seeds add flavor to gins and liqueurs. Roots are used in a tisane or tonic for treating fever, infections, and flu.

**Plant part used: Roots and stems**

### Myrrh *Commiphora myrrha* (NEES) ENGL.

Native to Ethiopia, Somalia, and the Arabian Peninsula, myrrh is prized for its fragrant reddish-brown or dark yellow oleo-gum-resin (combination of oil, gum, and resin) that exudes from bark openings. Myrrh has long been used for many medicinal purposes, including as an antibiotic, an antiseptic, and to enhance blood circulation.

**Plant part used: Gum (or resin)**

### Gum Arabic Tree *Acacia nilotica* (L.) DELILE

Gum arabic is the resin that exudes from *Acacia nilotica* and its close relatives in northern Africa, Arabia, and West Asia. It has seemingly endless uses as food, medicine, textiles, cosmetics, and more. Gum arabic is often used to improve certain qualities in wine.

**Plant part used:** Seeds and gum (or resin)

### Labrador *Ledum palustre* L.

Labrador, also called marsh tea or wild rosemary, is a beautiful rhododendron shrub from the tundras of Asia and North America. It contains poisonous compounds, but also therapeutic ones useful in treating rheumatism, microbial infections, and repelling insects; Linnaeus recorded peasants using it to ward off lice. It should always be water-extracted. It has a delicate flavor, resembling bay leaf.

**Plant part used:** Leaves

### Asafetida *Ferula assa-foetida* L.

"Fetida," from the Latin root for "fetid," is the first red flag for this powerful spice; another name is "devil's dung." The powder comes from rhizomes of a giant fennel plant and is used in cooking and as a digestive aid. Prepare in hot water and mix sparingly into bitters as a binder of diverse tastes, including sweet, salty, sour, and spicy.

**Plant part used:** Dried latex from the rhizome

### Valerian *Valeriana officinalis* L.

As an anxiolytic herb with sedative and analgesic effects, valerian is often used to treat insomnia. Beware of the valeric acid; no container can prevent this pungent root from overpowering your other herbs or home. Pair it with hops, chamomile, and lemon balm, antispasmodic and antianxiety herbs with complementary flavor profiles.

**Plant part used:** Roots and rhizomes

### Hyssop *Hyssopus officinalis* L.

An aromatic, slightly bitter member of the mint family, hyssop is often an ingredient of absinthe and vermouth. The first scientific description of penicillin was made from a sample from a decayed hyssop plant by the Swedish mycologist Westling. Hyssop contains large quantities of diosmin, a compound that can improve varicose veins.

**Plant part used:** Leaves and stalk

**Blutwurz** (or erect cinquefoil) *Potentilla erecta* USPENSKI EX LEDEB.

This small, upright plant grows wild in the mountains, heaths, and meadows of Europe, Western Asia, and Scandinavia. It is unique in the rose family for having only four petals, each with a distinctive notch. The bitter roots are the source of a red dye. Clinical studies show *Potentilla erecta* to be effective in treating ulcerative colitis.

**Plant part used:** Rhizome

# Asia

**Pepper** *Piper nigrum* L.

Black pepper originated in southwestern India and remains an important spice with an annual production of 80,000 tons (72,000 metric tons). It is used widely in traditional medicine to improve digestion and remove toxins, and it is used in Ayurvedic medicine to help relieve respiratory congestion. It should be consumed in small amounts to avoid irritations.

**Plant part used:** Seed

**Mace and Nutmeg** *Myristica fragrans* HOUTT.

Mace is the dry, red lacy flesh wrapping the nutmeg seed. Widely used since ancient times, the Spice Islands were named for these species. The oils have many beneficial functions, including soothing the stomach, lowering blood pressure, and as a liver tonic. Please note that the spices are often sold adulterated or with the oil already extracted.

**Plant part used:** Seed

**Cinnamon** *Cinnamomum cassia* (L.) J. PRESL, *Cinnamomum verum* J. PRESL, and *Cinnamomum loureiroi* NEES

Derived from the inner bark of evergreen trees in the genus *Cinnamomum*, the flavor derives from a range of volatile compounds, with cinnamaldehyde being the most distinct. The two main types of cinnamon are *C. cassia*, native to southern China, and *C. verum*, native to Sri Lanka and southern India.

**Plant part used:** Bark

**Cardamom** (black and green) *Amomum subulatum* ROXB. and *Elettaria cardamomum* (L.) MATON

Cardamom is one of the most distinct ingredients of chai and has an intense resinous fragrance. Derived from several species native to India, Bhutan, Indonesia, and Nepal, the black and green seeds vary in their flavor. Black cardamom has a smoky aroma with hints of coolness reminiscent of mint; green cardamom generally has a stronger flavor.

**Plant part used:** Seed

**Clove** *Syzygium aromaticum* (L.) MERR. & L. M. PERRY

Cloves come from the unopened flower buds of trees that grow up to 40 feet (12 m) and must be picked by hand before the flowers open. Wild clove stands can still be found in the North Molucca islands, where the species is native. Cloves are important in many cuisines and drinks, as a dental painkiller, and for a host of therapeutic uses.

**Plant part used:** Flower Buds

**Tea** *Camellia sinensis* (L.) KUNTZE

The most widely consumed drink in the world besides water, tea is a very diverse species with more than 1,500 cultivars and countless landraces. Tea is an elegant plant that can grow up to 60 feet (18 m) in its native form, and it has an extensive history of use as a medicine, tonic, beverage, and food for energy and well-being.

**Plant part used:** Leaves

**Black Caraway** (or black cumin) *Nigella sativa* L.

Nigella, known by many names, is used in Europe, northern Africa, India, and Southeast Asia. The seeds are sprinkled on breads; a potent, black seed oil can be applied topically or ingested. Nigella was cultivated early in Egypt and seeds were found in Tutankhamun's tomb. The seed are said to produce warmth and increase energy.

**Plant part used:** Seed

**Ginger** *Zingiber officinale* ROSCOE

Arab traders introduced ginger to the ancient Greeks and Romans, and today it is cultivated globally as an ornamental, culinary, and medicinal crop. For centuries, ginger has been vital in Chinese, Ayurvedic, and other medicines for treating catarrh, toothache, asthma, and more. It is not yet certain if ginger is safe during pregnancy.

**Plant part used:** Rhizome

### Galangal *Kaempferia galanga* L.

Native to Indonesia, southern China, Taiwan, Cambodia, and India, galangal looks and tastes similar to the rhizomes of ginger. This tropical species has long been used for similar purposes as ginger, as an herbal medicine and culinary spice, and the essential oil contains UV-absorptive properties found in natural sunscreens.

**Plant part used:** Rhizome

### Basil *Ocimum basilicum* L.

The leaves are used predominantly, but the seeds can be added to bitter drinks or desserts. The different cultivars have aromas that are often indicated by their names, including lemon, lime, and licorice basil. Basil is used around the world for rituals of death; in ancient Greece and Egypt, basil was placed in the hands of the dead to ensure a safe journey to the afterlife.

**Plant part used:** Leaves

### Apple *Malus domestica* BORKH.

Apples originated in mountainous central Asia. Today's cultivated apple is a complex hybrid from several species. The seedlings of apples, like many open-pollinated fruit trees, bear little resemblance to their parent plant. To use in bitters, keep the peel on but remove the seeds, which contain trace amounts of cyanide.

**Plant part used:** Fruit

### Star Anise *Illicium verum* HOOK. F.

Star anise comes from the dried dehiscent (those that split when mature) fruit pods of an evergreen tree native to Vietnam and China. Similar to cloves, the fruit is hand harvested, then dried. Use caution! Star anise can easily be mistaken for the smaller, irregularly shaped Japanese anise, or "shikimi" (*Ilicium anisatum*), which are poisonous.

**Plant part used:** Seed

### Turmeric *Curcuma longa* L.

One of the most heavily relied-on plants for its medicinal and culinary qualities, turmeric is integral to Chinese, Ayurvedic, Siddha, and Unani medicine. While it stains badly, light is all that is needed to completely fade the color. Neither turmeric pigment nor flavor extracts well in water, but results are better in alcohol or with an emulsifier.

**Plant part used:** Rhizome

### Peach *Prunus persica* (L.) BATSCH

Peaches and almonds are native to Asia and can breed with each other, despite diverging from a common ancestor eight million years ago. Both species were domesticated early in the first wave of tree-crop cultivation that swept the globe six thousand years ago. Peach pits, along with apricots, are used to make almond-flavored extracts.

**Plant part used:** Fruit

### Jasmine *Jasminum sambac* (L.) AITON

A fragrant species native to the eastern Himalayas in Bhutan and neighboring India and Pakistan, jasmine adds scent to ceremonial garlands and crowns. In China, the flower is used to flavor jasmine tea. It has been used for its health attributes, including detoxification, as a mood improver, and to support healthy blood flow.

**Plant part used:** Leaves

### Jujube *Ziziphus jujuba* MILL.

The jujube (pronounced joo-joo-bee) fruit grows on a spiny, shrublike tree native to China, with a present natural distribution to southeastern Europe, and its use in food and medicine dates back more than four thousand years. Packed with minerals and vitamin C, jujube is an important nutritional food in China. Jujube candies, also called jujyfruits, originally contained jujube juice, used as a cough remedy.

**Plant part used:** Fruit

### Mume Plum *Prunus mume* SIEBOLD & ZUCC.

With more than three hundred cultivars of the mume plum with differing flower colors, it is perhaps the most important ornamental tree in East Asia. Mume plums are widely used in liquor, juices, and sauces and as a flavoring. The pits can produce hydrogen cyanide and should not be casually eaten.

**Plant part used:** Fruit

### Indian Gooseberry (or amla) *Phyllanthus emblica* L.

Amla is primarily sour and astringent, secondarily sweet, bitter, and pungent, giving it one of the broadest taste profiles, lacking only salt. The Hindu god Vishnu is said to dwell in amla (gooseberry) trees, and many traditions, including Ayurveda, indicate its importance as a cure-all, life extender, and tool for achieving enlightenment.

**Plant part used:** Fruit

### Areca Nut *Areca catechu* L.

The nut of the fruit of the areca palm is chewed along with betel leaf throughout India, the Philippines, China, Taiwan, Vietnam, Malaysia, Myanmar, and the Pacific. While the extract has been shown to have antidepressant properties, it must be used sparingly, because it can be addictive and excessive use has direct links to oral cancers.

**Plant part used:** Seed

### Schisandra *Schisandra chinensis* (TURCZ.) BAILL.

Schisandra is a deciduous woody vine native to the forests of northeastern China, Japan, and Korea. The fruit has a remarkable flavor profile, with all five basic tastes recognized in traditional Chinese medicine (see page 38). Among other benefits, this plant is rich in antioxidants, but it should not be used during pregnancy or breastfeeding unless supervised by a qualified health-care provider.

**Plant part used:** Fruit

### Monk Fruit *Siraitia grosvenorii* (SWINGLE) C. JEFFREY EX A. M. LU & ZHI Y. ZHANG

Like its relative, pumpkin, monk fruit grows as twining vines and is cultivated along trellises. For centuries, it has been used in food and beverages as a natural sweetener and as a traditional Chinese medicine for the treatment of pulmonary ailments and dry coughs, among other conditions. Its Chinese name is *luo han guo*.

**Plant part used:** Fruit

### Banaba *Lagerstroemia speciosa* (L.) PERS.

Native to southern tropical Asia, banaba trees are now widespread. They have clusters of thin, crepe-textured flowers. They do not provide nectar, but the trees help maintain healthy bee populations by producing two types of pollen: one for reproduction and one for bee food. Fresh or dried leaves, bark, roots, and fruit can all be used.

**Plant part used:** Leaves

### Pandan Leaf *Pandanus amaryllifolius* ROXB.

The mild, distinct nutty taste of pandan is remarkable for its ability to transform the flavor of rice dishes. Pandan is delicious when steeped in coconut milk, and it also serves multiple therapeutic roles when prepared as bitters, including as a digestive aid and to reduce fever.

**Plant part used:** Leaves

## Tulsi *Ocimum tenuiflorum* L.

Also called "holy basil," tulsi originated in north-central India and its distribution corresponds with cultural migrations. According to folklore, it grew from the tears of Vishnu. Use with caution for anyone with diabetes or hypoglycemia. Be careful when mixing with other species believed to affect blood sugar, such as monk fruit.

**Plant part used:** Leaves

## Kousa Dogwood *Cornus kousa* BÜRGER EX HANCE

Recognizable by what appears to be large four-petaled flowers, this species, and other dogwoods, are ornamental trees from Asia that are now common across temperate and subtropical regions of the Old and New worlds. Both the fruit (fresh and dried) and leaves are edible, but we have not found the latter in beverage recipes.

**Plant part used:** Fruit

## Barley *Hordeum vulgare* L.

Barley's use in fermentation for beer and as fodder for animals is as ancient as its use as food. Recipes on Mesopotamian clay tablets document how beer was made from bread, not sprouted grain as in modern brewing. In Afghanistan, Egypt, and South Korea, the flowers and fruit are used as a traditional contraceptive.

**Plant part used:** Seed

## Bael *Aegle marmelos* (L.) CORRÊA

The leathery pulp can be strained to make refreshing cocktails and mocktails, and the fresh leaves can be eaten as salad greens. The bael tree is widespread in India, Thailand, Myanmar, and other Southeast Asian countries. Regarded as sacred in Hinduism, it is part of a fertility ritual for girls in Hindu and Buddhist traditions in Nepal.

**Plant part used:** Fruit

## Blue Pea (or Asian pigeonwings) *Clitoria ternatea* L.

This colorful, evocative flower, straight out of a Georgia O'Keefe's painting, is one of our favorite ingredients to work with. The root has a sharp bitter taste with anti-inflammatory properties, but it is used as an emetic and should not be consumed by women while pregnant. The flavor of the flower extract is light, sweet, and somewhat vegetal.

**Plant part used:** Flower

# Africa

### Tamarind *Tamarindus indica* L.

Native to tropical Africa, all parts of tamarind trees are used heavily. The leaves and flowers are sour and floral and are used as famine foods and in curries; the roots and bark are used to treat abdominal pain. We encourage mixtures of water and ethanol for extractions; they are shown to diminish diabetes symptoms and cholesterol.

**Plant part used:** Fruit

### Coffee *Coffea* sp.

Originally prepared for spiritual purposes more than one thousand years ago, the pulp of fermented coffee berries was prepared and used during religious ceremonies. The earliest evidence of the coffee drinking that we are more with familiar today, from roasted and brewed coffee beans, is from the fifteenth century. Coffee is widely known for its stimulant properties.

**Plant part used:** Seed

### Grains of Selim *Xylopia aethiopica* (DUNAL) A. RICH.

The grains of Selim are used to spice coffee to make the spiritual drink café Touba, named after a holy city in Senegal. Originating around Ethiopia and now produced widely in Ghana, the flavor is densely aromatic, musky, resinous, and pungent. In several traditional African pharmacopoeias, it is used to aid the healing of the uterus after giving birth.

**Plant part used:** Seed

### Kola Nut *Cola* spp.

Two species are used interchangeably to produce a stimulant flavoring known as "cola," initially a primary ingredient of Coca Cola. It is native to west Africa, where it is chewed as a social lubricant and stimulant, and has a major religious importance in Nigeria. Nuts are boiled to extract the flavor—bitter and astringent when fresh, sweet and roselike when dried.

**Plant part used:** Fruit

### Baobab *Adansonia digitata* L.

The baobab is the iconic tree of the African bush, and its relationship with humans is entwined with folklore and religion. Its habitat ranges from Mauritania and Sudan in the north to South Africa. All parts are used. The fruit and seeds are rich in vitamin C, but nearly all parts can be eaten; some are used for medicine, dyes, or to make goods.

**Plant part used:** Seed and fruit

### Grains of Paradise *Aframomum melegueta* (ROSCOE) K. SCHUM.

From ancient times, grains of paradise have been traded for their intense woody aroma, transported through the Sahara to destinations throughout Italy and other parts of Europe, where they became popular in the fourteenth and fifteenth centuries. Today, grains of paradise continue to be used to flavor some craft beers and gins.

**Plant part used:** Seeds

### Honeybush *Cyclopia* spp.

The first recorded use of this tea tree was in the 1770s, and the consumption of honeybush tea has now become global practice. Decoctions of the plant have traditionally been used in the Cape region as a restorative. Recently, powdered extracts from fermented honeybush have been used in the cosmetic industry.

**Plant part used:** Leaves, stems, and flowers

### Ajwain *Trachyspermum ammi* (L.) SPRAGUE

Ajwain is a branched annual herb. It is a highly valued, medicinally important seed spice and also used as a flavoring and condiment. In traditional Ayurvedic medicine, ajwain is primarily used for stomach disorders. The crushed fruit is used topically as a poultice, and the seeds can be used for their aphrodisiac properties.

**Plant part used:** Leaves and fruit

# Europe and the Near East

**Strawberry** *Fragaria* x *ananassa* (WESTON) DUCHESNE EX ROZIER

The French were cultivating *Fragaria vesca*, the woodland strawberry, in the fourteenth century, but a breakthrough occurred in 1714, when a French spy introduced the Chilean strawberry and the two hybridized naturally. Strawberry is an accessory fruit, meaning that the sweet, edible part is an enlarged fleshy receptacle, so the seeds are on the outside. Strawberries have a large amount of DNA—so much that you can easily isolate it at home.

**Plant part used: Fruit**

**Common Wormwood** (or mugwort) *Artemisia vulgaris* L.

Mugwort and its relatives are used across the northern hemisphere for their green, piney, herbaceous flavoring. It is used in food and medicine as a digestive stimulant and a lightly mind-bending tea. May Day is also known as Mugwort Day in Romania, where mugwort adorns the home and is carried to cast out evil spirits.

**Plant part used: Leaves and stalks**

**Juniper** *Juniperus communis* L.

In addition to gin, whose name derives from *genevre*, the Old French word meaning "juniper," the astringent blue-black seed cones are also used to flavor traditional farmhouse ales of Norway, Sweden, Finland, and Estonia. Valued as a medicine and an herbal tea in North America, one of its uses has been as a contraceptive by the Navajo people.

**Plant part used: Seed cone (commonly called berries) and leaves**

**Hops** *Humulus lupulus* L.

What do hipsters and a twelfth-century saint have in common? The IPA. The first mention of the preservative qualities of hops in beer came from Saint Hildegard von Bingen's writings as an abbess; she even alluded to their sedative properties. Hops are traditionally used for treating tension and anxiety. Studies have shown that they improve sleep, especially when partnered with valerian.

**Plant part used: Flowers**

### Elderberry *Sambucus nigra* L.

Elderberries and flowers are common ingredients in North Africa, Europe, and North America. The flowers express a strong, floral, sweet tropical flavor; the dominant compounds are soluble in warm alcohol but not water. The berries are always cooked, usually boiled, to destroy toxic compounds.

**Plant part used:** Fruit and flowers

### Chicory *Cichorium intybus* L.

Chicory is native to Europe, where it is found growing on roadsides. The leaves of this herbaceous species are used as bitter greens for salads, and the roots are often used as a coffee substitute. Chicory, with its coffeelike notes, is used as a flavoring for stouts and for blond Belgian-style ales to augment hops.

**Plant part used:** Roots

### Pomegranate *Punica granatum* L.

The pomegranate has numerous significances in various cultures globally and all major religions. Frequently, it is used to symbolize wealth, health, prosperity, and fertility. Pomegranate has been used medicinally for thousands of years, ingested and used topically for conditions ranging from heart problems, dysentery, and intestinal parasites to oral conditions and wrinkles.

**Plant part used:** Fruit

### Mullein *Verbascum thapsus* L.

The many names of mullein, including the descriptive "cowboy's toilet paper," pay homage to its dense, fuzzy leaves. Mullein leaves have many uses dating back millennia, particularly in the treatment of respiratory ailments. Innocuous and mild in flavor, you can mix it into nearly any blend with honey or propolis to combat a cough or cold.

**Plant part used:** Leaves

### Fennel Seed *Foeniculum vulgare* MILL.

Fennel originated in the southern Mediterranean and is indispensible in regional cooking. Both the stalk and fruit, known as the seed, have valuable nutritive qualities and are used for everything from breath freshener to pain relief. The plant has naturalized across many parts of the world and has become an invasive weed in numerous places.

**Plant part used:** Seed

### Lavender *Lavandula stoechas* L. and *Lavandula angustifolia* MILL.

Native to the Mediterranean, lavender has a versatile, deep floral, and spicy aroma with nuances of mint and lemon. It pairs well with sweet and savory flavors. A little goes a long way in cocktails and foods; too much results in a soapy, waxy taste. Make sure you use only food-grade culinary lavender.

**Plant part used:** Flowers and sometimes leaves and stalks

### Artichoke *Cynara scolymus* L.

Thistles were a favorite of Eeyore, and we can share his appreciation through the unripe buds of a domesticated variety. In the sixteenth century, artichokes were thought to be an aphrodisiac and only men were allowed to consumed them to enhance libido. Artichoke-base bitters will quickly overpower; pair with other bold flavors.

**Plant part used:** Leaves

### Saffron *Crocus sativus* L.

One of the oldest, most valuable, and widespread multipurpose plants under cultivation, saffron has an ancient history of use as a pigment, medicine, and spice, with saffron-base pigments being found in fifty-thousand-year-old prehistoric cave art in Iraq. Throughout history, its value has been on a par with gold.

**Plant part used:** Stigma

### Bay Laurel *Laurus nobilis* L.

The Oracle of Delphi is said to have consumed or inhaled laurel leaves that put her in a trance, through which she could communicate with the god Apollo and provide cryptic advice. Unless you are allergic to the sesquiterpene lactones (a type of phytochemical compound) laurel produces, it is safe to consume and has many beneficial medical qualities, including treating asthma.

**Plant part used:** Leaves

### Caraway *Carum carvi* L.

Native to the eastern Mediterranean and North Africa, caraway is often used for its wide-ranging medicinal properties. In Europe, it was grown in medieval monasteries for its antiflatulent capacity. The fruit is used as a popular flavoring in liqueurs, savory cooking, and desserts, the seeds are widely used, and the roots and leaves can also be eaten.

**Plant part used:** Fruit

**Grape** *Vitis vinifera* L.

Today, grapes are cultivated on every continent except Antarctica, with vines growing best in temperate climates. Grapes have a long history of use for food, beverage, and medicine. The earliest human use of grapes as food was during the Neolithic period, and wine storage jars dating back seven thousand years have been found in Iran.

**Plant part used:** Fruit

# South America

**Quinine** *Cinchona* spp.

Quinine, derived from cinchona bark, was the standard treatment for malaria from the seventeenth century until the 1940s. The addition of sugar and carbonated water made the medicine taste better—and a dash of gin even made the tonic water pleasant. Its widespread associations with Jesuit missionaries and the Catholic Church led to its nicknames "Jesuit bark" and "Cardinal's bark."

**Plant part used:** Bark

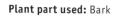

**Cacao** *Theobroma cacao* L.

Cacao—chocolate—is native to the upper Amazon and was domesticated by the Mayas. It was used traditionally to treat a range of conditions, but when it arrived in Europe in the late sixteenth century, chocolate was used to treat hot diseases, such as fevers. Its use eventually widened as it was tested within the European medical system of the time.

**Plant part used:** Seed

**Pineapple** *Ananas comosus* (L.) MERR.

Pineapple originated in the understories of the forests of southern Brazil and Paraguay. They are prized for their many health benefits—the most well known being bromelain, a powerful anti-inflammatory used for general pain relief. However, pineapple is also known to induce miscarriage, particularly the young fruit, which has high levels of bromelain.

**Plant part used:** Fruit

### Habanero Chili Pepper *Capsicum chinense* JACQ.

The habanero originated in South America and apparently reached Mexico from Cuba, hence the name. This fast-growing species forms one of the hottest chilies in the world. The fruit has been used as a tonic, an antiseptic, and, perhaps unsurprisingly, to increase perspiration. It is important to avoid breathing in chili pepper vapors or getting any extract in the eyes.

**Plant part used:** Fruit with seeds

### Quinoa Seed *Chenopodium quinoa* WILLD.

The seed coat of this annual herbaceous plant is rich in bitter saponins, soaplike foaming phytochemicals that have antioxidant properties but can be a mild digestive irritant. Rinsing and drying quinoa removes some of these compounds. Quinoa originated in the Andean regions of Peru, Bolivia, Ecuador, and Colombia, where it was domesticated three thousand to four thousand years ago.

**Plant part used:** Seeds

### Amaranth *Amaranthus* spp.

In parts of the world, notably China and Africa, leaves of *Amaranthus* species are widely used potherbs and a major source of dietary protein. The taste is mild and earthy, like oatmeal. Use the ornamental cultivars as a source of betalain to add color to your tincture or drink.

**Plant part used:** Seed

### Açaí *Euterpe oleracea* MART.

Before açaí fruit became marketed as a superfood for surfers it was mainly an oily staple in Brazil and Trinidad with impressive levels of protein and fiber. Some Amazonian communities relied on açaí fruit for nearly half their calories. The oils of açaí drupes (fruit) are sensitive to heat and can taste rancid. Use high-proof alcohol for extraction.

**Plant part used:** Fruit

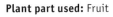

### Inca Berry *Physalis peruviana* L.

Inca berry, also known as ground cherry, Cape gooseberry, and physalis, tastes like the love child of tomato and pineapple and is native to the high altitudes of South America. The sweet-and-sour flavor of the antioxidant-rich fruit pairs well with lemon juice to bring out its sour notes and agave to accentuate its sweetness.

**Plant part used:** Fruit

## Maca *Lepidium meyenii* WALP.

Maca looks a little like a turnip—both are in the Brassicaceae family. Unlike turnip, however, the dried powder made from the Andean superfood has a pleasant, almost peanut-buttery taste. Maca has been used traditionally as an adaptogen and aphrodisiac, and it has been shown to improve semen quality both in healthy and infertile men.

**Plant part used:** Roots

## Muña *Minthostachys mollis* (KUNTH) GRISEB.

Muña stands out in the Andes for its cold-enduring green leaves, and is it important in the Peruvian pharmacopoeia. Commonly consumed as a tea or a minty inhalant, muña is also a spice for savory dishes and a preservative. Note that alcohol extracts obtained by soaking muña can be more bitter than the aqueous extracts or distillation.

**Plant part used:** Leaves

## Guarana *Paullinia cupana* KUNTH

Guarana has high caffeine levels but is generally taken in small quantities. The phytochemicals support metabolism, cognition, cardiovascular health, lower depression, and reduce inflammation. In the United States, fruit powder and seed extract are approved as flavorings, but do not have "generally recognized as safe" (GRAS) status from the Food and Drug Administration.

**Plant part used:** Seed and sometimes fruit pulp

## Hoja Santa (or Mexican pepperleaf) *Piper auritum* KUNTH

A member of the pepper family with heart-shape leaves that smell like root beer, hoja santa is revered in Mexico, where the name translates to "sacred leaf." The name may stem from the legend that Mary hung Jesus's diapers to dry on the plant, which took on the sweet scent. It is an ingredient in the regional liqueur verdín, named for its green color.

**Plant part used:** Leaves

## Guayusa *Ilex guayusa* LOES.

An evergreen tree in the *Holly* genus, guayusa has been cultivated by tribes, particularly in eastern Ecuador, to use as an energizing drink that targets the central nervous system. Water extracts are earthy and slightly sweet, not bitter. You can make a concentrated tea; mix it with soda, liquor, and honey syrup to make an energizing cocktail.

**Plant part used:** Leaves

### Pau D'Arco *Handroanthus impetiginosus* (MART. EX DC.) MATTOS

The inner bark of pau d'arco is prepared by several South American indigenous peoples as a bitter tisane, used as a general tonic, an immune stimulant, anti-inflammatory, adaptogen aid to fight candida, and promote lung health. The hardwood species with gorgeous pink flowers was used for making hunting bows.

**Plant part used:** Bark

### Lúcuma *Pouteria lucuma* (RUIZ & PAV.) KUNTZE

The earliest appearance of lúcuma dates from 5,700 to 3,000 BCE Peru, which today is the main source with more than 85 percent of global production. Lúcuma has a sweet taste although it is low in sugar; the flesh of the fruit is often processed as a flavoring for ice cream, bakery products, yogurt, and more.

**Plant part used:** Fruit

### Damiana *Turnera diffusa* WILLD. EX SCHULT.

The Guaycura, a tribe in Baja, were believed to be the first to use damiana in religious ceremonies, and they traded it with the Aztecs as a potent aphrodisiac. Distilleries in southern Baja still use damiana in liqueurs. It is believed to have been an original ingredient of margaritas before triple sec. Damiana is pungent, with figlike notes and a bitter aftertaste.

**Plant part used:** Leaves

# Mexico, Central America, and the Caribbean

### Allspice *Pimenta dioica* (L.) MERR.

Reportedly used in embalming practices by the Mayas, allspice is often used today in health and beauty products, including toothpaste flavoring. Medicinally, allspice may have immune-stimulating effects and is applied topically for muscle pain and toothache. Allspice in large amounts might interact negatively with anticoagulant medicines.

**Plant part used:** Fruit

## Cascarilla *Croton eluteria* (L.) W. WRIGHT

The bark of cascarilla is used in herbal medicine and as an aromatic bitter for its function as a tonic, stimulant, and fever reducer; it is also used as a flavoring in Campari and vermouth. Studies show the bark can tone the stomach to help improve its function and increase appetite and can strengthen the histamine-stimulated gastric acid secretion.

**Plant part used:** Bark

## Epazote *Dysphania ambrosioides* (L.) MOSYAKIN & CLEMANTS

The pungent, acidic, lemony flavor of epazote is unique. The leaves of this wild herb, native to Mexico and the United States, have long been added to flatulence-inducing beans in Mexican cooking to reduce their effects and as an overall aid to digestion, making it an excellent ingredient for digestive bitters.

**Plant part used:** Leaves

## Prickly Pear *Opuntia ficus-indica* (L.) MILL

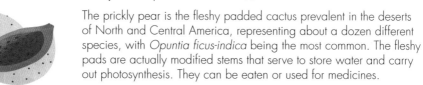

The prickly pear is the fleshy padded cactus prevalent in the deserts of North and Central America, representing about a dozen different species, with *Opuntia ficus-indica* being the most common. The fleshy pads are actually modified stems that serve to store water and carry out photosynthesis. They can be eaten or used for medicines.

**Plant part used:** Fruit

## Sarsaparilla *Smilax ornata* LEM.

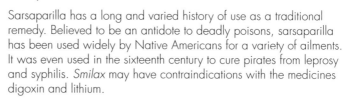

Sarsaparilla has a long and varied history of use as a traditional remedy. Believed to be an antidote to deadly poisons, sarsaparilla has been used widely by Native Americans for a variety of ailments. It was even used in the sixteenth century to cure pirates from leprosy and syphilis. *Smilax* may have contraindications with the medicines digoxin and lithium.

**Plant part used:** Roots

## Vanilla *Vanilla* spp.

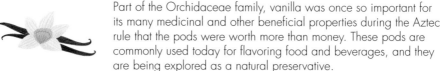

Part of the Orchidaceae family, vanilla was once so important for its many medicinal and other beneficial properties during the Aztec rule that the pods were worth more than money. These pods are commonly used today for flavoring food and beverages, and they are being explored as a natural preservative.

**Plant part used:** Seed

### Gordolobo *Gnaphalium* spp.

The yellow flowers of this soft, woolly biennial are prepared in a tisane for supporting digestion and gastrointestinal issues. It has also been used as a topical remedy for skin, throat, and breathing ailments. The Aztecs used gordolobo for treating coughs, sore throats, diarrhea, and dysentery.

**Plant part used: Leaves and flowers**

### Devil's Hand *Chiranthodendron pentadactylon* LARREAT.

The flowers look like blood-red creepy claws and are difficult to acquire. Apparently, the Aztecs celebrated the tree in Mexico, where there were only a few specimens, but they grow in abundance next door in Guatemala. The flowers don't have much flavor, but they tell a photogenic story about climate change and conservation.

**Plant part used: Flowers**

### Avocado *Persea americana* MILL.

Native to south Central America, avocados have been used since the earliest human habitation, with evidence dating back twelve thousand years. Different parts of the tree are used in a variety of traditional remedies. Avocado leaf and bark extracts should be avoided during pregnancy and all of the plant, except the fruit pulp, is toxic to livestock and fish.

**Plant part used: Fruit**

# North America

### Sugar Maple *Acer saccharum* MARSHALL

Sugar maple—the sap of which is used to produce maple syrup—is a long-lived species that does not reach reproductive maturity until at least twenty-two years old. A tea from the inner bark has been used as a diuretic, infusions are used to treat eye problems, and maple syrup is often used in cough syrups.

**Plant part used: Sap**

## Cranberry *Vaccinium macrocarpon* AITON

A cousin of the blueberry, cranberry is a North American native. Native Americans introduced it to early settlers (hence its association with Thanksgiving). The fruit of the cranberry plant, which is extremely tart and acidic, has a long history of use for food and medicine in North America and Europe; it is used to treat urinary tract infection.

**Plant part used:** Fruit

## Wintergreen *Gaultheria procumbens* L.

The sweet, spicy leaves of wintergreen inspired synthetic flavoring of breath mints, chewing gum, and toothpaste. Used by several Native American tribes for medicinal purposes, the fruit, also known as teaberry, is edible with mild notes of sweet wintergreen; the leaves and branches are dried and infused into a tisane.

**Plant part used:** Leaves

## Blueberry *Vaccinium corymbosum* L. and other *Vaccinium* spp.

Blueberries were important in the diet of indigenous peoples in North America, who believed that the Great Spirit had sent them from the stars to feed them and maintain health, hence marking the fruit with a star at the bottom. Excessive consumption of blueberry leaf tea can be toxic and should be avoided, especially with diabetes medicines.

**Plant part used:** Fruit

## Prickly Ash *Zanthoxylum clava-herculis* L.

The fruit and bark have been traditionally used as a stimulant and tonic for digestive organs. However, the bark produces an acrid and astringent mouthfeel, leading to copious salivation and a numbing effect similar to Sichuan pepper; use it sparingly in blends. Prickly ash may have contraindications with blood-clotting medicines.

**Plant part used:** Leaves

## Osha *Ligusticum porteri* J. M. COULT. & ROSE

Osha is a perennial herb in the carrot family. Known as bear medicine, osha holds a strong association with bears in spiritual ceremonies of the First Nations people of North America. The roots have been used in various preparations internally for colds and other respiratory ailments and externally to treat pain. Avoid osha during pregnancy.

**Plant part used:** Roots

# Australia

### Lemon Myrtle *Backhousia citriodora* F. MUELL.

Extremely versatile, lemon myrtle has been used fresh, dried, or as an essential oil as a bush food in Australia, in medicines, and in flavorings. During World War II, beverage company Tarax used the essence as a flavoring when their preferred lemon essence was scarce. Medicinally, the oil is believed to be sanitizing and have antifungal and antimicrobial compounds.

**Plant part used:** Leaves

### Gum Trees *Eucalyptus* spp.

Thought to total 70 percent of trees in Australia, eucalyptus was imported to the United States and widely planted during the American Gold Rush. Australian indigenous medicine uses it in many ways, including the smoke from burning leaves for flu and fever, and modern medicine is investigating if it has anticancer properties. Taking higher than recommended doses of eucalyptus oil can be toxic.

**Plant part used:** Leaves

### Australian Mint Bush *Prostanthera rotundifolia* R. BR.

Also called Australian native thyme, the plant is native to the eastern coast of Australia, where the leaf is used as a food and medicine. It has purported antioxidant and anticancer properties and has ancient use as a medicinal inhalant, clears sinuses, improves breathing, and is a carminative. It is now largely used for food and beverages.

**Plant part used:** Leaves and stalks

### Mountain Pepper/Pepperberry *Tasmannia lanceolata* (POIR.) A. C. SM.

The woody, herbal flavor of mountain pepper echoes allspice and cinnamon. Study of the plant morphology has identified special oil cells in the leaves that are used in defense, releasing the toxic compound only when the leaf is physically damaged. You can make extracts of the leaves or berries in vinegar or grain alcohol.

**Plant part used:** leaves and fruit

### Native Australian Ginger *Alpinia caerulea* (R.BR.) BENTH.

The white pulp of this uncommon ginger is native to Australia.
A perennial herb growing under wet forest canopy, it is prized for
its mild sweet, spicy, and sour flavors, which activate the salivary
glands. It is used by Aboriginal peoples to moisten their mouths
during bushwalking in arid conditions.

**Plant part used:** Rhizome

### Paperbark Tree *Melaleuca leucadendra* (L.) L.

Many *Melaleuca* species are nicknamed "tea tree"—although
they are more closely related to guava than the actual tea plant.
A passable substitute for tea is made from an infusion of the young
leaves. In Australia, the bark was used as a body wrap in Aboriginal
burial ceremonies. It has natural antiseptic properties, and a cooling
and pungent flavor.

**Plant part used:** Leaves, flowers, and nectar

### Desert Raisin *Solanum centrale* J. M. BLACK

The highly aromatic fruit, rich in vitamin C, of this thorny "bush tomato",
a member of the deadly nightshade family, has been a source of food
for Aboriginal people for thousands of years. The dried ground fruit is
added to bread mixes, salads, sauces, chutneys, stews, and butter.

**Plant part used:** Fruit

### Yams *Dioscorea* spp.

The human relationship with yams began in West Africa, where
they could be relied upon for survival. Today, the many culinary and
medicinal uses include being a treatment for cramps, coughs, and
hiccups. Many species are inedible; others have tubers that must
be boiled or soaked to remove toxins. Yam leaves must be cooked
before consumption.

**Plant part used:** Fruit

# Making Your Own Extractions

Although the Shoots & Roots recipes (pages 52–67) will help you make our own house bitters, we hope that one of the reasons you are reading this book is because you want to experiment to make your own bitters recipes using the same method and your own choice of botanicals and solvent. However, it's vital to observe some basic safety procedures to make sure that the process and product are safe.

### Choosing Your Plants

Select safe and sustainably sourced botanicals. Make sure the botanicals are not toxic for humans, and keep in mind that traditional use does not guarantee a botanical's safety; some plants are nontoxic as water extracts but can be harmful as alcoholic extracts. In those cases, you can make a concentrated water extract and add it to alcohol later on.

Search reputable academic papers on ethanol and methanol extracts on all the species you use. If you come across anything worrying, investigate it farther.

In preparing bitters with safety in mind, we take the following approach.

01. We focus on the safety of botanicals in the preparation of our bitters; we can make no guarantees that they will be effective as medicines and do not suggest people make bitters to self-medicate. Always first consult a physician if you are taking medication.

02. We only use botanicals that have a long history of common use in traditional cultures and then cross-check their safety through databases and scientific literature.

03. We do not use the following botanicals commonly used in traditional home remedies for preparing alcoholic bitters. Please note that there may be other botanicals not in this list that are commonly used but aren't safe for bitters.

- Arnica (*Arnica* spp.)
- Creosote/chaparral (*Larrea tridentata*)
- Ginkgo leaf (*Ginkgo biloba*)
- Kava kava (*Piper methysticum*)
- St. John's wort (*Hypericum perforatum*)
- Birthwort (*Aristolochia* spp.)
- Digitalis (*Digitalis* spp.)
- Madder root (*Rubia tinctorum*)
- Senna (*Senna alata*)
- Wood fern (*Dryopteris* spp.)

## Choosing Your Solvents

Some phytochemicals are more soluble than others; as a rule of thumb, the less soluble the chemicals, the higher proof of alcohol they need. At Shoots & Roots, we generally start to process with the highest proof grain alcohol that we can find and prepare a first extraction, then follow this with a water extraction, using water that has been ultrafiltered or freshly boiled to eliminate bacteria.

The ratio of plant to solvent is also a key factor. Working in metric, we have developed a ratio of 50 grams of dried plant material to $1\frac{5}{8}$ cups (375 mL) of solvent. Please note that this represents the total botanical material, so all the individual species in a recipe add up to a total of 50 grams.

Some bitters recipes call for different types of alcohol that have a lower proof, including wine, hard cider, and vodka. These lower proof alcohol types are useful when working with plants that have a large amount of tannins; that keeps the extraction levels down and prevents that overwhelming astringent mouthfeel. Bitters recipes with lower-proof types of alcohol don't need to be diluted with a water extract or after the alcohol extraction.

## Nonalcoholic Solvents

You can use vegetable glycerine to make nonalcoholic glycerite bitters. Replace alcohol with 75 percent glycerin and 25 percent water, then follow the basic method. When you finish filtering your bitters in step 6, you can go directly to bottling your bitters in step 9, skipping steps 7 and 8.

## Getting Your ABV Right

Does the bitters you have created need to be diluted to a different ABV (alcohol by volume)? When diluting, you will want to consider stability. To create a shelf-stable product, the ethanol content should be high enough to be considered bacteriolytic (above 33 percent), meaning it will kill bacteria. We standardize out bitters to 42 percent ABV. If the blend has fruit content that is whole or can settle on its own, we standardize to 45–50 percent ABV. Use the following formula in diluting to your desired concentration:

$$\text{Volume of Water to Add} = \frac{\text{Solvent ABV} \times \text{Solvent Volume}}{\text{Desired ABV}} - \text{Alcohol Extract Volume}$$

For example, to dilute $1\frac{3}{4}$ cups (415 mL) of an extract made using $1\frac{5}{8}$ cups (375 mL) of 75 percent ABV (150-proof) grain alcohol to a final ABV of 45 percent, add $\frac{7}{8}$ cup plus 2 teaspoons (215 mL) of water:

$$\text{Volume of Water to Add} = \frac{75 \times 375 \text{ mL}}{45} - 415 \text{ mL}$$

$$\text{Volume of Water to Add} = 625 \text{ mL} - 415 \text{ mL}$$

You can mix the bitters and test how close to the desired ABV you are by using your hydrometer, a tool for testing alcohol content. If you are low, add a little more alcohol extract. If you are high, add a little more water extract. The liquid should be at room temperature for more accurate readings. Add sweeteners only after you balance the ethanol content: sugar or glycerin make the mix viscous, which will throw off the hydrometer reading.

## Notes for Extracting Different Phytochemicals and Adjusting ABV

| Type of Phytochemical | Solubility in Water/Extraction Notes |
|---|---|
| **Alkaloids** | Generally require high-proof alcohol for extraction (there are exceptions, such as caffeine) |
| **Flavonoids** | Generally extract well in water and low-to-moderate-proof alcohol (40 proof) |
| **Organic acids** | Generally extract well in water and low-proof alcohol |
| **Lactones** | Generally require high-proof alcohol for extraction |
| **Saponins** | Generally extract well in water and low-proof alcohol |
| **Sugars** | Extract well in water and low-proof alcohol |
| **Tannins** | Generally extract well in water and low-proof alcohol for a short period of time. Overextracting tannins results in an astringent mouthfeel. |
| **Terpenes** | Generally require high-proof alcohol |

| Solvent Volume | Solvent ABV | Desired ABV | Final Volume | Volume of Water to Add |
|---|---|---|---|---|
| 1⅝ cups (375 mL) | 95% (190 proof) | 45% (90 proof) | 3⅜ cups (792 mL) | 792 mL minus Solvent Volume |
| 1⅝ cups (375 mL) | 95% (190 proof) | 42% (84 proof) | 3⅝ cups (848 mL) | 848 mL minus Solvent Volume |
| 1⅝ cups (375 mL) | 75% (150 proof) | 45% (90 proof) | 2⅝ cups plus 2 tsp (625 mL) | 625 mL minus Solvent Volume |
| 1⅝ cups (375 mL) | 75% (150 proof) | 42% (84 proof) | 2⅞ cups plus 1 tsp (669 mL) | 669 mL minus Solvent Volume |

# Using a Sonicator

As scientists, one of the unique processes that we use for making Shoots & Roots Bitters draws from our phytochemical lab work in preparing botanical extractions in a process known as sonication. Sonication applies sound energy to extract phytochemicals from plant material into the liquid used for making bitters. This results in a faster extraction and often enhances the amounts of phytochemicals that we are able to draw out of a plant.

When making botanical extractions in the lab, we place our bottled-up mix of botanical-and-solvent mixtures into a sonicator (also known as an ultrasonicator and ultrasonic cleaner) in order optimize the extraction process. Sonicators are essentially water baths that use ultrasonic sound frequencies to create jets of pressure that serve to agitate botanicals including disrupting their cell walls. This results in a quicker and increased release of the phytochemicals from the botanical material into the solvent. At Shoots & Roots Bitters, we have adapted this process in making bitters. If you are looking to add a laboratory process to your kitchen or bar for more optimal extractions, a sonicator is a relatively accessible option.

We add the sonication step after we shake up our botanical extract (Step 4 of the Bitters Masterclass on page 61). If using sonication, fill up the ultrasonicator with cold water and place the lidded jar with your botanical-and-solvent mixture in the water bath for 30 minutes at 86°F (30°C). As sonication relies on sound rather thermal dynamics, it is a mild extraction method that prevents degradation of the active phytochemicals. However, as temperature of the water bath does increase during sonication, be mindful to switch out the water for cold water if making multiple batches of bitters. Warmer temperatures of the water bath can increase extraction capacity, but they can also degrade certain compounds, such as vitamins C and E. You can most often see how well the sonication works through a quick color change of the solvent—often this is comparable to waiting several days to a few weeks of storing botanical-and-solvent mixtures.

# Blending Flavors: Pairing and Flavor Matching

**Bitters require pairing certain flavors to get the best drink. Think of a base layer with bitter or earthy notes, a middle layer of citrus or spice, and then a top layer of the fragrance, which may be floral, warming, and so on. It's important to compose bitters with ingredients that spread across all three layers to make a drink more complex and enjoyable to the senses. The trick is to balance them so one layer doesn't dominate, unless that is your intention.**

### Matching Flavors to a Liquor

It helps to think about the flavor of the liquor, which usually goes into the medium and high layers of flavor and aroma, so you don't compete with but complement it when adding bitters. The bitters should be thought of as the medium to carry the scent of other ingredients as well as bitters ingredients. If you add vermouth that has a lot of fruity notes, you can work with bitters that have a lot of rich spice, bitterness, and depth. Bitters work well to balance the sugar content in cocktails and that includes the liquor base. If making a drink for the first time, keep the bitters light and add them gradually.

Note that the temperature and texture of the drink affects how much you can detect the bitters. Tonic or anything carbonated masks the base layers but brings out the top layer, and it makes drinks seem less sweet by the nature of carbon dioxide being sour. A lot of ice slows down the movement of bitters into the air, so you can be a more liberal than if working with hot drinks, especially if the bitters are being shaken into the cocktail instead of being added as a topper.

If making a flip or adding egg white, the proteins keep the bitters from jumping into the air above your beverage. They hold the bitters in parts of the glass, so they are wonderful for making a pretty design. A heavy dose of bitters adds to the middle notes that mask the egg flavor, which comes out as the drink warms up.

We made bitters that we thought were interesting, then sought people to design drinks around them. There are endless possibilities for making drinks and for making the bitters themselves. Most common bitters recipes include gentian or extraordinary bitter root. We encourage you to use a broader range of plants for the bitter base and other layers and to seek versatility.

# Resources

## Ingredient suppliers

Bulk Apothecary
www.bulkapothecary.com
Frontier Co-op
www.frontiercoop.com
HerbalCom
www.herbalcom.com
Herb Wholesalers
www.herbwholesalers.com
Kalustyan's
foodsofnations.com
Mountain Rose Herbs
www.mountainroseherbs.com
Monterey Bay Spice Company
www.herbco.com
Monterey Botanicals
www.montereybotanicals.com
Oregon's Wild Harvest
www.oregonswildharvest.com
Pacific Botanicals
www.pacificbotanicals.com
SF Herb Co.
www.sfherb.com
The Teaspot
theteaspot.com
Golden Bough Botanicals Inc.
goldenbough.ca
Canadian Herbalist's Association
of BC Supplier List
www.chaofbc.ca

## Websites

American Botanical Council
cms.herbalgram.org/BAP
TRC Natural Medicines Research
Collaboration
naturalmedicines.
therapeuticresearch.com

## Books

*Agaves of Continental North America* by Howard Scott Gentry
*Beachbum Berry's Potions of the Caribbean* by Jeff Berry
*Bitters* by Brad Thomas Parsons
*Compendium of Materia Medic* by Li Shizhen
*Crops and Man* by Jack Harlan
*Curating Biocultural Collections*
by Jan Salick, Katie Konchar, and Mark Nesbitt
*Darwin's Harvest*
by Timothy Motley, Nyree Zerega, and Hugh Cross
*Domestication of Plants in the Old World*
by Daniel Zohary, Maria Hopf, and Ehud Weiss
*The Drunken Botanist* by Amy Stewart
*Duke's Handbook of Medicinal Plants of the Bible*
by James Duke
*Jones' Complete Bar Guide* by Stan Jones
*Liquid Intelligence* by Dave Arnold
*Lost Crops of Africa* (3 volumes) by National Research Council
*Medicinal Plants of the Philippines* by T. H. Pardo de Tavera
*The Murder of Nikolai Vavilov* by Peter Pringle
*Native American Food Plants: An Ethnobotanical Dictionary*
by Daniel Moerman
*One River* by Wade Davis
*Shattering: Food, Politics and the Loss of Genetic Diversity*
Cary Fowler and Pat Mooney
*Tales of a Shaman's Apprentice* by Mark J. Plotkin
*Tea Horse Road: China's Ancient Trade Road to Tibet*
by Michael Freeman and Selena Ahmed
*The Tequila Ambassador* by Tomas Estes
*Vintage Spirits and Forgotten Cocktails* by Ted Haigh
*WD-50: The Cookbook* by Wylie Dufresne and Peter Meehan

## Journals

Economic Botany
Ethnobiology Letters
Human Ecology
Journal of Ethnobiology and Ethnomedicine
Journal of Ethnopharmacology

# Glossary

**adaptation:** Genetic change over generations that improves the survival of a population in a particular context such as a new habitat.

**adaptogen:** A substance that when taken helps one cope with stress, usually by strenghtening the body's ability to combat stress. Ginseng is a common example.

**alchemical:** Historic traditions or trying to perfect certain objects, often associated with the ancient effort to create gold from other elements, which proved impossible.

**analgesic:** Acting as a pain reliever.

**anaphylaxis:** An acute allergic reaction.

**angina:** Chest pain caused by not enough oxygen-rich blood getting to the heart.

**angioedema:** A condition of swelling beneath the skin.

**anticarcinogenic:** Having the abitily to fight cancer-causing carcinogens.

**anticoagulant:** Having the ability to slow or reduce the coagulation of blood.

**antidote:** A medicine that counteracts a poison, sometimes called a reversal agent.

**antiflatulent:** A substance with the ability to minimize flatulence.

**antihyperlipidemic:** A promoter of lower lipids, such as cholesterol, in the bloodstream.

**antihypertensive:** Having the ability to lower high blood pressure.

**antinociceptive:** Having the ability to reduce sensitivity to pain.

**antinutritionals:** Substances that reduce the nutritional availability in foods, beverages, and fodders.

**antioxidants:** Substances that inhibit oxidation of molecules, thereby reducing their production of cancer-causing free radicals.

**antiplatelet:** Having the ability to prevent platelet aggregation, reducing susceptibility to blood clots and blocked blood vessels.

**antiproliferative:** Having the effect of limiting cell growth, such as tumor growth.

**anxiolytic:** Substances with anti-anxiety properties.

**areoles:** Small bumps from which spines emerge on cacti that are actually specialized branches.

**bacteriostatic:** Having the ability to prevent bacteria from reproducing or changing behavior without killing them.

**bacteriolytic:** Having the ability to destroy bacteria.

**bonsai:** A Japanese art form in which trees are cultivated in small containers, often made from cuttings of larger trees.

**calabash:** A hollowed bottle gourd used as a storage vessel or utensil.

**candida:** A fungus that occurs naturally on the human body, but under certain environmental conditions grows out of control, causing yeast infections.

**carminatives:** Substances that reduce bloating and flatulence.

**catechin:** A type of flavanol with astringent taste and antioxidant properties that is common in tea.

**cochineal:** Often used to describe the reddish color source of carmine, derived from the female cochineal beetle.

**contraindications:** Situations where a substance may be harmful to a person, perhaps because of interactions with another drug.

**diaphoretic:** Having the ability to induce or increase sweating.

**dysentery:** A condition caused by any of several bacteria, viruses, protists, or worms, that causes severe diarhhea with blood.

**dyspepsia:** Upper abdominal indigestion with symptoms of pain, heartburn, belching, and more.

**embalming:** Methods to preserve deceased organisms from decay and often to laden with fragrance.

**emetic:** A substance that causes vomiting.

**expectorant:** A substance that increases the water in mucus so it can be loosened and coughed out.

**hypocotyl:** The stem of a germinating plant, between the cotyledons and the root.

**hypoglycemia:** A condition of having low blood sugar, which can lead to a variety of symptoms from shakiness to seizures.

**Levant:** The eastern Mediterranean, extending into parts of Syria, Saudi Arabia, and Egypt, where many crops were domesticated.

**nightcap:** A beverage intended to be ingested before bed, to promote relaxation and rest.

**vasodilation:** The widening of blood vessels that in turn reduces blood pressure.

**xanthones:** A type of health-beneficial compound usually rare in nature but prevalent in mangosteen.

# Index

# Acknowledgments

We are grateful to the people, plants, and landscapes that have inspired and nurtured us. This book has been influenced by hundreds of thousands of people over centuries and their uses of botanicals that are ultimately the foundation of this work. We are honored and humbled to work with communities around the world who have shared their ecological knowledge and cultural practices while giving us a window into indigenous lifeways. Explorers and scientists throughout history have inspired us to pursue research questions as well as to share science with the world through unconventional means. This book has further been inspired by gastronomic innovations and creative forces in bars and kitchens in our backyards.

We are indebted to the tireless editors at Ivy Press who have wonderfully shaped this book, including Jenny Campbell, Tom Kitch, Imogen Palmer, and Monica Perdoni. Illustrator Clare Owen and photographers Xavier Buendia and Stephen DeVries added vibrancy to our text with their visual contribution. We are grateful for the devotion of the team at Roost Books to this book project.

Our adventures with Shoots & Roots Bitters would not have been possible without the support of our families, partners, friends, colleagues, mentors, collaborators, customers, workshop participants, and event attendees. Christian Schaal and Jim Merson shared their sacred cocktail and soda recipes featured in Chapter Five. Our Shoots & Roots Advisory board, including Kevin Denton, Jesse Falowitz, Jennifer Fistere, and Michael Purugganan, have provided invaluable feedback on our bitters creations, and our approach.

We have benefited incredibly from the sharing of time, ideas, beverage making, and event collaboration by the following people and organizations: Adam Schmidt, Amy Bessa, Anna Herforth, Ariela Zycherman, Beth Denton, Bittercube, Botanical Society of America, Brian Matthys, Christina Gittings, Coco Redfield, Daniel Kulakowski, Daria Mazey, Edward Kennelly, Eleanor Friend, Ellen Arnstein, Emeric Harney, Eric Gardner, Eunsuk Bae, Gul Ahmed, Harlem Commune, Harney and Sons Tea, HerbalGram, John Kao, John Munafo, Jon van Deren, Joren MacMillan, Kenyatta Bell, Kurt Reynerston, Lou Sagar, Louis Putzel, Marcus Samuelsson, Mark Bitterman, Mark Blumenthal, Martim Ake Smith-Mattsson, Mary Kelly, Maya Edelman, Michael Harney, Michele Gazzolo, Michelle Olsgaard Stewart, Mitch Harris, Nada Ahmed, Natalia Pabon-Mora, Nate Dorr, Nazli Ali, Noah ten Broek, Noel, Oded Brenner, Patricia Wichmann, Rachel Sussman, Rick Visser, Runa Tea, Sam Brissette, Samuel Barickman, Scott Pancake, Shayon Chatterjee, Society for Economic Botany, SoHyun Kim, Stephanie Wang, Stigma Members and Organizers, Strawberry Hill Grand Delights, The Gin Garden, Trudy Chan, Uchenna Unachukwu, World Maker's Faire Organizers, and Zàhra Ahmed.

We have tremendous gratitude to Hot Bread Kitchen in Harlem, New York, and their HBK Incubates program that brought Shoots & Roots Bitters into operation and shared their commercial kitchen space. We are inspired by Hot Bread Kitchen's model in paying it forward by supporting small entrepreneurs to mitigate start-up risk and grow their food ventures. In similar vein, we thank the team at Kickstarter and the supporters of our Kickstarter campaign, who have the spirit to see the full potential in people and their ideas.

A number of academic institutions and their faculty trained us to be the botanists and scientists we are today including The New York Botanical Garden, The City University of New York Plant Sciences Doctoral Program, The School of Anthropology and Conservation at the University of Kent, The New York University Center for Genomics and Systems Biology, Kew Gardens, Yale School of Forestry and Environmental Studies, Department of Horticultural Sciences at the University of Florida at Gainesville, The U.S. Botanic Garden, Barnard College, Columbia University, Tufts University, Arnold Arboretum at Harvard University, University of California at Berkeley, and the Fairchild Tropical Botanic Garden.

This book would not exist without your support. Thank you

**Selena Ahmed, Ashley DuVal, and Rachel Meyer.**